Sustainable Con

What is a *durable* economy? It is one that not only survives but thrives. How is it created, and what does it take to sustain it with a high quality-of-life outcomes? *Sustainable Communities* provides insight and answers to these questions.

Citing Burlington, Vermont's remarkable rise to award-winning status, this book explores the balance of community planning, social enterprise development, energy and environment, food systems and cultural well-being. Aimed at policymakers, development practitioners, students, and citizens, this book describes which and how multiple influences facilitate the creation of a local, durable and truly sustainable economy. The authors hope to inspire others by sharing this story of what can be done in the name of community economic development.

Rhonda Phillips Community well-being is the focus of Rhonda's research and outreach activities. Author/editor of 15 books on community development and related topics, she offers both practice and academic perspectives on the ever changing topic of community revitalization as a professor at Arizona State University's School of Community Resources and Development.

Bruce Seifer is a consultant with deep experience in economic development. He led the City of Burlington, Vermont's economic development efforts for three decades, providing technical assistance to 4,000 businesses and numerous nonprofits. Bruce frequently speaks at national forums on policy and strategy, city revitalization, and program design and evaluation.

Ed Antczak After 20 years in business, Ed joined CEDO's Economic Development Division in 2003 focusing on assisting businesses at all stages of growth, managing a revolving loan fund, and being a member of various development project teams. He currently serves on the steering committees of several national sustainable economic development organizations.

'Burlington is an amazing city, with innovations that make the mind race. How did they do all this? Their successes in creating one of the country's most economically viable cities isn't born of luck, nor is it the work of socialist, ice-cream eating hippies. It is the product of hard work, focus and a fierce commitment to the power of community.

The authors have penned a book that should be read by everybody who is working for a new economic vision for America. College students, social entrepreneurs, nonprofit leaders, city administrators and every elected leader in America. Our economic recovery won't be achieved by big projects or multinational businesses. The future isn't in that direction. As Burlington proves, it's as easy as looking in your own backyard.'

Robert Egger, President LA Kitchen & CForward www.lakitchen.org

'This is a truly invaluable book! Building upon the exemplary experience of Burlington, Vermont it offers real guidance on how to achieve sustainability, community wealth-building, worker-ownership, and democratic control. A must read for practioners, scholars and students concerned with rebuilding the American system from the ground up on strong principles anchored in strong communities.'

Gar Alperovitz, Lionel R. Bauman Professor of Political Economy at the University of Maryland, and author, What Then Must We Do?

'In an era of a dysfunctional federal government, Sustainable Communities: Creating a Durable Local Economy reminds us that change at the local level is possible and that we can all be the better for it. Phillips, Seifer and Antczak document a remarkable adventure set in Burlington, Vermont which demonstrates how indigenous leadership coupled with a strong sense of community can make a difference. They have a written a playbook for transformational change that matters and which can be sustained.'

Nicolas P. Retsinas, Director Emeritus,
Harvard University's Joint Center for Housing Studies

'Rhonda Phillips, Bruce Seifer, and Ed Antczak have written about sustainable economic and community development from a great point of view. They have actually done it.

Based on their extensive experience in Burlington, Vermont, a city of 40,000 in Northwestern New England with a long history of commercial and industrial development, the three authors have put together a terrific combination of theory and practice in writing about this popular and elusive goal. This is a scholarly work and a how to manual wrapped together, and it has a happy ending. Burlington is fast becoming one of the most sustainable and livable cities on the planet.'

Howard Dean, former Vermont Governor 1991-2003

'Sustainable Communities represents a groundswell of local initiatives; a bottom-up alternative that recaptures the creativity of civic spirit, and the melding of practice and vision. Vermont and Burlington in particular have been blessed with leaders committed to building a community that works for all, and does it with a lively spirit. Bruce Seifer has spent decades as one of those leaders, and Sustainable Communities: Creating a Durable Local Economy delivers on its promise, offering inspiring and practical lessons on what works, the application to other towns and cities, and even some implications for the nation.

Paul Freundlich, Founder and President Emeritus, "Green/Co-op America"

"This book tells the story of one of our best models of a sustainable and desirable community in the US. We desperately need these models and we even more desperately need to spread the word about how it happened and what we can learn from it. The world is approaching a tipping point and with the help of models and books like this we can help it tip into something much better."

Robert Costanza Chair in Public Policy, Crawford School of Public Policy, The Australian National University, Canberra

'We are in DEEP SHIFT! Sustainable Communities is an essential and practical guide to building resilient, equitable and lively communities that will be well positioned for the economic and ecological changes ahead. The Burlington VT story is inspiring and replicable.'

Chuck Collins, Institute for Policy Studies and the Resilience Circle network (localcircles.org), Boston, MA

'A beautifully organized how-to-book demonstrating the many ways municipal governments can work hand-in-hand with business and engaged citizens to build a local economy that is green, fair and fun. This real-life working model offers an invaluable tool to the localist movement.'

Judy Wicks, Co-founder of BALLE and author, Good Morning, Beautiful Business.

'If you want to actually know how to build a great and vibrant community, here's the down-and-dirty details from the people who've done it. Absolutely essential!'

Bill McKibben, author, Oil and Honey: the Education of an Unlikely Activist

'Sustainable Communities is more than a case study or how to guide to make your own 'durable local economy' for those already engaged in community development.

The City of Burlington, VT itself is an inspiring story. Perched on the northern shores of Lake Champlain, it has become a beacon for those looking for fresh ideas, new business models, and the keys to creative, fun neighborhoods. The real take away for all of us who care about the places where we live, work and play is this: it all starts with, revolves around, and depends on people. Sustainable communities engage people at all levels and sectors - social, cultural, environmental, political. I have seen it work in my own backyard, Bath, ME, recognized as one of the "Best Small Cities in America". Read this book; you may be inspired to unleash your own inner 'civic entrepreneur' and contribute to making where you live a more durable, sustainable place.'

Eloise Vitelli, Arrowsic, Maine (Just across the river from Bath)-Director of Program and Policy Development, Maine Centers for Women, Work, and Community

'This informative and detailed book provides insight, examples, and, best of all, hope – about the possibilities of creating a strong, equitable and inclusive local economy. Well-written with many cases, examples and practical information, the authors draw on years of experience to show how to build an economy that works for the private sector, the public sector and residents of the community. The book will certainly appeal to community and economic development practitioners both within and outside the United States.'

Michael Swack, Faculty Director, Center on Social Innovation and Finance, The Carsey Institute, University of New Hampshire

'For 30 years, Burlington Vermont has been intentionally making its economy stronger and more resilient. Here is the detailed back-story: rigorous attention to existing local employers, affirmative entrepreneurship aid to single mothers and immigrants, employee ownership assistance, import substitution, cultivation of environmental resources, creation and preservation of farmland and local farm-consumer linkages, birthing two dozen specialized non-profits, promotion of arts and culture, post-industrial place-making, strategic and sparing use of incentives—and marketing the resulting successes! Every local official or activist seeking inspiration should explore this book.'

Greg LeRoy, Executive Director, Good Jobs First, Washington, DC

'Burlington Vermont may be the best model we have for the Next Economy (the one we need to replace the one we've got, which just isn't working, plain and simple, for our planet and most of the people on it). This is a wonderful story about combining enlightened community development, strong civic partnerships, and 30 years of hard work to assemble the components of a restorative economy for the future.'

John Abrams, President and CEO, South Mountain Co., Inc. and author, Companies We Keep

'Burlington is one of those places that offers both opportunity and inspiration to its residents and businesses. This book provides insight into how community inclusion and vision became business as usual in this small city...with remarkable results.'

Jerry Greenfield, Co-Founder, Ben and Jerry's

'This is a must read for every public official, especially town managers, mayors, administrators and select men and women who want a view of the possible in leading positive change in their communities, creating livable jobs, and using their powers as elected or appointed representatives to offer the Municipality writ large as an organizing catalyst to important economic, social and environmental outcomes as if people mattered.

Historically, community-based economic development and CDCs (community development corporations) arose in the civil rights era of the 1960s as an antidote to the failure of municipal government to serve as a locus of economic opportunity and enfranchisement of people and communities out of the economic mainstream. The authors' book turns this fact of history on its head with a real life historical documentary of the Municipality of Burlington, Vermont's record of practice and policy of community-based economic development.

Springing from sound economics and redefinition of a return on investment to include both the social and environmental benefits to be gained, whether from mainstreet small business shops featuring locally-made goods and foods, affordable housing, or high quality and environmentally-sound manufacturing ventures, the book chronicles the important work of the town's Community and Economic Development Office (CEDO). Its cast of public officials led the town into a variety of partnerships with the private as well as nonprofit third sector and now the so-called fourth sector of social enterprises, yielding a picture of how a municipality can help build sustainable, healthy and "resilient" communities – that is, communities that can bounce back if the people are engaged.'

Ron Phillips, president, CEI ~Capital for Opportunity & Change, Wiscasset, Maine

'For more than a decade I've admired the authors and their leadership in Burlington. Theirs is an example of a community being the change that's needed. And now I'm so grateful for this book, which lays out for all of us, a framework and the essence of what it takes to build a durable local economy.'

Michelle Long, Executive Director, BALLE: Be a Localist

Earthscan Tools for Community Planning Series

Sustainable Communities by **Rhonda Phillips, Bruce Seifer and Ed Antczak**
July 2013 | Paperback 978-0-415-82017-2 | Hardback 978-0-415-82016-5 | Ebook 978-0-203-38121-2

The Placemaker's Guide to Building Community by **Nabeel Hamdi**
April 2010 | Paperback 978-1-84407-803-5 | Hardback 978-1-84407-802-8 | Ebook 978-1-84977-517-5

Creative Community Planning by **Wendy Sarkissian, Dianna Hurford and Christine Wenman**
February 2010 | Paperback 978-1-84407-703-8 | Hardback 978-1-84407-846-2 | Ebook 978-1-84977-473-4

Kitchen Table Sustainability by **Wendy Sarkissian, Nancy Hofer, Yollanda Shore, Steph Vajda and Cathy Wilkinson**
November 2008 | Paperback 978-1-84407-614-7 | Ebook 978-1-84977-179-5

The Community Planning Event Manual by **Nick Wates**
August 2008 | Paperback 978-1-84407-492-1 | Ebook 978-1-84977-293-8

The Community Planning Handbook by **Nick Wates**
October 1999 | Paperback 978-1-85383-654-1 | Ebook 978-1-84977-600-4

Forthcoming:
The Community Planning Handbook, Second Edition by **Nick Wates**
October 2013 | Paperback 978-1-84407-490-7

Sustainable Communities

Creating a Durable Local Economy

Rhonda Phillips, Bruce Seifer and Ed Antczak

First published 2013
by Routledge
2 Park Square, Milton Park, Abingdon, Oxon, OX14 4RN

Simultaneously published in the USA and Canada
by Routledge
711 Third Avenue, New York, NY 10017

Routledge is an imprint of the Taylor & Francis Group, an informa business

British Library Cataloguing-in-Publication Data
A catalogue record for this book is available from the British Library

Library of Congress Cataloging in Publication Data
A catalog record has been requested for this book

ISBN13: 978-0-415-82016-5 (hbk)
ISBN13: 978-0-415-82017-2 (pbk)
ISBN13: 978-0-203-38121-2 (ebk)

Typeset in Frutiger and Stone Serif by
Saxon Graphics Ltd, Derby

Printed and bound in Great Britain by
TJ International Ltd, Padstow, Cornwall

Contents

Illustrations

Figures

Tables

Foreword

Creating the Durable Economy: How Burlington Became an Award-winning City

Senator Bernie Sanders

Community development, done right, might just be one of the more important things we can do to save America.

Today, Congress too often turns its back on the pressing needs of state and local governments, and on the basic aspirations of most Americans who want a decent, secure job to provide for their families and their future, and a sustainable and thriving community in which they can live.

Today, most Americans are becoming more and more dissatisfied with our misplaced national priorities.

They see trillions of dollars going to defense spending and seemingly unending foreign conflicts. They see outrageous tax breaks for millionaires and billionaires during a deep and painful recession. They see state and local governments facing huge budget deficits, unable to maintain basic services such as police protection or education, to say nothing of finding funds to create new jobs, provide better health care, protect the environment, and fix our decaying roads, bridges, and water and sewer systems.

Our nation is crying out for help at the local level, but the richest, most powerful people and corporations continue to demand more and more for themselves while they demand that our nation shift more and more costs – and cuts to essential programs – onto the shoulders of working American families.

The crooks on Wall Street whose greed, recklessness and illegal actions precipitated this recession are now earning more money than before the American people bailed them out. In America today we have the most unequal distribution of income and wealth of any major country on earth, and the gap between the very rich and everyone else is growing wider. Today, the top 1 percent earns more income than the bottom 50 percent of Americans. In 2010, 93 percent of all new income went to just the top 1 percent. While "official" unemployment is at 8.1 percent, real unemployment is over 15 percent – and is especially high for blue-collar workers who have been the backbone of many local economies.

America's working families are losing hope that government can help to make their lives better. Too often they feel they must opt out of the political and civic process to concentrate on basic survival.

Community development projects, large or small, have the potential to bring people out of their despair and cynicism and give them a stake in civic life and hope for the future. Many community development projects are truly "seeds" – small, effective, tenacious organisms that can endure a lot of stress, and which possess enormous potential given their starting size.

Many of the answers to America's troubles, I believe, will be found largely at the state and local level, in creative, inclusive economic development policies and actions that answer local needs and involve ordinary citizens in the process.

This book tells the story of how innovative, and often courageous, economic development projects pushed a small city on the path to serve all of its citizens.

There are many good lessons that can be learned from listening to the people who guided these efforts. I don't necessarily wish for readers to copy Burlington's actions step by step, as the way government, small businesses, and non-profit organizations work together will vary from city to city or state to state.

But I do hope that people will absorb what I think is the book's most important, and proven, message – that when you work to meet the needs of all residents, a city will shine in a special way, and will become an inviting place where others will want to visit, live, and do business.

There is no question that America is at a critical point today. Our middle class is disappearing, poverty is increasing, more and more young people cannot find work, and the gap between the very rich and everyone else is growing wider.

Now is the time for ordinary people in cities and towns across the country, in hundreds of different community development efforts, to become leaders in the fight to create a better America, to create new jobs and a renewed sense of community.

Bernie Sanders
U.S. Senator
Washington, D.C.
October 2012

Preface

It is tricky telling a compelling story about creating and maintaining a functioning local economy and community, especially when creating a narrative about an economy and community that strives to take care of the vulnerable, and builds hope because the model is working – something of an anomaly in this era of economic downturn. These perspectives are offered as a way to chronicle the remarkable journey of a city at the northern reaches of the U.S. border. Burlington, Vermont is one of those unique places that can't be ignored; it's different and in that difference is something special – a palpable, indomitable community spirit that has helped shape an award-winning city with a durable and resilient economy. It's remarkable because so many notables have been achieved, all in one small city. It's been nearly 30 years since Burlington embarked on a new approach to community and economic development, with the mission of fostering economic vitality; preserving and enhancing neighborhoods, quality of life, and the environment; and promoting equity and opportunity for all residents. Many have worked diligently towards these goals, accomplishing much, suffering some setbacks, and receiving quite a few accolades along the way. Burlington is often referred to as a model of how an engaged municipal government and citizenry can play an active role in helping create and foster a healthy and vibrant local economy. And for Burlington, it's clearly not about development for development's sake. It is about actualizing a vision of what people want in their community.

What is a durable economy? It is essentially a resilient economy that perseveres – not only surviving, but thriving. It is comprised of many components or sectors that we have termed the following: the locally-focused economy sector, the social economy sector, the cultural economy sector, the environmentally sustainable economy sector, the political economy sector, and the inclusive economy sector. All communities have these sectors, or variations of them, to varying degrees in their economies. Taking these components or sectors together has led to the formation of a durable economy in Burlington and they are presented as a framework for understanding how this was accomplished. Just focusing on the traditional aspects of an economy isn't enough. Without addressing all the sectors, there is less of a chance that a durable economy will form. Burlington is a laboratory, and understanding how each of these sectors works together may be instructive for other communities in efforts to build their own durable economy. It is hoped that this story will inspire, or at least intrigue, others to explore what can be done in the name of community economic development.

What's in store: how this book is organized

Chapter 1 sets the framework for the rest of the book, introducing how Burlington's trek towards a durable economy and award-winning status began, and why it continues today. Also described is the tipping point for Burlington occurring in the early 1980s, and how the community has maintained a sense of purpose and commitment to being the unique place it is. Efforts centering on fostering durability in the economy, by building long-term capacity and focused action, are introduced. The stage is also set with a discussion of what some call a fourth sector mindset – how organizations and their leaders are transforming communities with social entrepreneurship and civic leadership. This first chapter focuses on an important component of a durable economy – the political economy sector – and how that influences so many other aspects of community economic development.

Each of the next six chapters focuses on a major area that has influenced and shaped Burlington through the years – locally-focused business development and socially responsible businesses, economic inclusiveness, cultural vibrancy, social well-being, energy and the environment, and local food systems. Each chapter begins with an overarching question, an overview and context of the topic, followed by an exploration of Burlington's approaches and outcomes. At the end of every chapter is a "Closer Look" section, providing a case illustrating results in each of the topical areas. All chapters conclude with a resource list providing links to websites and other sources of helpful information, including documents and sites referenced in each chapter as well as links to national and global resources.

Chapter 2 centers on the locally-focused economy sector with the power of local – how focusing on locally driven development and locally owned businesses provides a foundation for building economic development strategies. Since 1984, locally-focused enterprise development has been a priority in Burlington. This involves employee-owned enterprises, buy local campaigns, and a living wage ordinance. It's also about fostering capacity, including fourth sector organizations such as socially responsible businesses. Our Closer Look case selection centers on Burlington's Local Ownership Development initiatives.

In Chapter 3, the inclusive economy sector shows the importance of including all in the economy, detailing how to expand women's entrepreneurial skills and reach out to the refugee community via small business development. The Closer Look case selection is about the Vermont Refugee Microenterprise Program. Next, Chapter 4 presents the cultural economy sector so vital in fostering community and economic development outcomes, looking at the transformation of a dying industrial area into an arts and creative business incubator district. The nonprofit organization, the South End Arts and Business Association, is our Closer Look selection.

Chapter 5 outlines the social economy sector, focusing on social well-being, and includes a case that has garnered national media attention, the Good News Garage. Energy and environmental issues are central to the economic development story in Burlington and Chapter 6 provides the background for both, centering on the environmentally sustainable economy sector. The Closer Look case is about the Vermont Sustainable Jobs Fund.

Chapter 7 illuminates another vital component of the environmentally sustainable economy sector in a community – food systems. The Intervale Center, a 350-acre "breadbasket" and model of urban agriculture is our Closer Look selection. At the end of the book, Chapter 8 sums up lessons learned, and outlines suggestions and considerations for fostering a durable economy. The final Closer Look selection is emblematic of what it takes to accomplish community economic development, with the multiple-year chronicle of City Market, a brownfield site revitalization success story of the creation of a downtown grocery cooperative. It is hoped that the information presented in this book will be useful to other communities in their economic development efforts, drawing together ideas and insights into how to foster a more durable local economy.

Acknowledgments

First of all, this book is a tribute to the many dedicated partners, citizens, leaders, and other civic and social entrepreneurs who over the last several decades have worked diligently to create and maintain Burlington's resiliency as a high-quality, dynamic place to live, work, and play. It is truly a community that made a durable economy happen.

There are so many who helped make this book a reality and we are grateful to each and every one. We want to thank and recognize our intern extraordinaire, Alison Flint, who was a tremendous asset to this project. Special thanks are due to our community partners, leaders, business owners, and citizens who gave graciously and patiently of their time and insights during interviews. These include: Jim Lampman, Bill Mitchell, Tom Longstreth, Michael Monte, Doug Hoffer, Brenden Keleher, Pat Robbins, Hal Colston, Peter Clavelle, Robbie Harold, Ben Cohen, Jerry Greenfield, Martha Whitney, Bill Truex, Betsy Ferries, Paul Bruhn, Alan Newman, Jim Flint, Mark Stephenson, Diana Carminati, Wayne Fawbush, Melinda Moulton, Will Raap, Ernie Pomerleau, Steve Conant, Beth Sachs, Leigh Steele, Rachel Hooper, Gyllian Rae Svensson, Yiota Ahladas, Travis Marcotte, Roy Feldman, and Clem Nilan. Finally, reams of appreciation are due to Robbie Harold, Richard Schramm, Julie Davis, Nancy Brooks, Robert Leaver and Hollis Hope for their excellent editorial advice and review.

In memory of Bentley Davis Seifer
who lives on in the heart of a community
1998–2011

1 The quest for a durable local economy

How Can Economic Durability and Community Resiliency Be Encouraged?

Burlington, Vermont

"Top 10 City for the Next Decade" – *Kiplinger's Personal Finance*, June 2010.

"#1 Happiest Small City in U.S., March 2011, *Gallup*."[1]

"10th lowest unemployment rate in the nation" – *U.S. Bureau of Labor Statistics*, July, 2012.

2nd lowest foreclosure rate in the U.S., 2010 – *RealtyTrac*, March, 2011.

Community resiliency is the topic of the moment, given current economic conditions. It implies the ability to be resilient in the face of challenges. From the Latin word, *resallive*, it literally means to spring back. It also has deeper meaning; the Community and Regional Resilience Institute points out the following characteristics of resiliency:[2]

- Attribute: resilience is an attribute of the community.
- Continuing: a community's resilience is an inherent and dynamic part of the community.
- Adaptation: the community can adapt to adversity.
- Trajectory: adaptation leads to a positive outcome for the community relative to its state after the crisis, especially in terms of its functionality.

While these characteristics are applied mostly in the context of responses to environmental disasters, there's relevance for community and economic development at the local level too.

Building resiliency in a local economy and society is a dynamic process, and requires continuing effort, as well as adaptation to changing conditions (and certainly some of those can be considered adverse!). It's also about wanting positive outcomes for the community, with planning and action aimed at this trajectory. As the Stockholm Resilience Center defines it, "resilience is the capacity of a system to continually change and adapt yet remain within critical thresholds."[3]

The title of this book reflects both durability and resilience, which imply being able to withstand the test of time. Bill McKibben, in *Deep Economy: The Wealth of Communities and the Durable Future* issues a call for an economy that creates and supports community and ennobles lives: "For the first time in human history, 'more' is no longer synonymous with 'better' – indeed, for many of us, they have become almost opposites." McKibben puts forward a new way to think about the things people buy, the food they eat, the energy used, and the money that pays for it all. "Our purchases," he says, "need not be at odds with the things we truly value."[4] Inspiration is found in this work, and it is noted that durability is both a noble goal and essential requirement for healthy local economies. His work also inspires an exploration of the full spectrum of an economy, incorporating social, cultural, political, inclusive, and environmentally sustainable sectors or components contributing to a more durable, resilient economy.

And what about these sectors or components of an economy as introduced in the preface? It's been the experience of the authors that it takes many types of efforts from the public, private, and nonprofit arenas to foster durability and resilience over time. It requires a combination of economic components and a host of cooperative stakeholders. Figure 1.1 illustrates the interconnection of these components, and it's no coincidence it's a lifesaver pattern – to stay afloat in any economy requires many sectors working together.

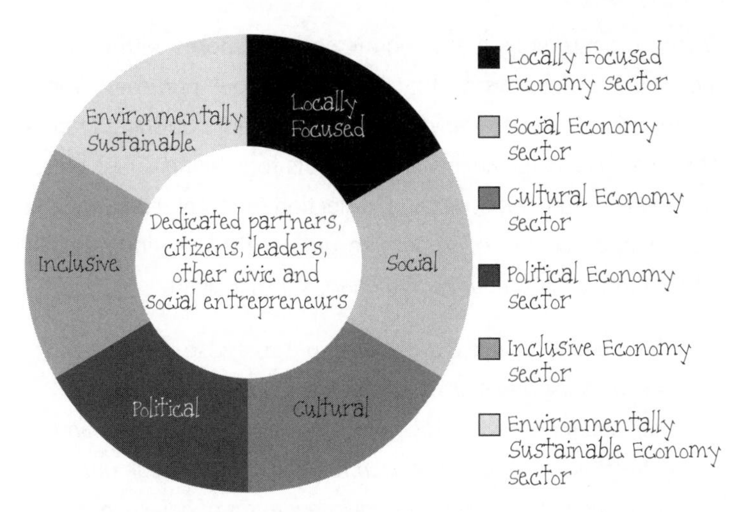

Figure 1.1 Creating a durable local economy

There is a direct connection between resiliency and building capacity at the community level, and the way to foster this connection begins with a mindset of social and civic entrepreneurship. It's all about fostering community for mutual benefit. Social and civic entrepreneurs are described as those who help:

> *Communities collaborate to develop and organize their economic assets and to build productive, resilient relationships across the public, private, and civil sectors. They forge the ties that bind economy and community for their mutual benefit. They provide continuity to work on tough issues over the long haul. The term civic entrepreneur combines two important traditions: entrepreneurship – the spirit of enterprise – and civic virtue – the spirit of community. Social entrepreneurs are change agents.*[5]

Much is happening in this arena now – witness the terminology emerging: mission-related investing, social enterprise, common good corporations, green businesses and sustainable enterprises, corporate social responsibility, social marketing, and economic sustainability. And it's happening in what some now call the *Fourth*

Sector: organizations that integrate social purposes with business methods,[6] or businesses that integrate social purposes. These organizations can be in any sector with the distinguishing feature of serving as change agents and catalysts for making things better (including communities). It's a reflection, too, of creating and sustaining a "culture of something greater than the individual."[7] John Abrams, in *Companies We Keep*, states it clearly:

> *We need to change our definition of success, so that it's less about doing and making as much as we can as fast as we can and more about satisfying human needs as elegantly and effectively as we can. We need to think about enough rather than more. We need to consider new forms of governance and business.*[8]

These new forms of governance and business hold transformative potential for our communities, regions and nations.

Economic development, as if people mattered

All of the types of components, or sectors, discussed in this book (locally-focused, social, cultural, environmentally sustainable, inclusive, and political) relate to economic development and creating a durable economy. Actually, it more than relates. It is akin to being foundational, or essential, and works both ways – economic development both as an innovative approach to any of these types while any of these sectors serve as a gateway to launch innovative economic development initiatives. Either way, the focus is on human and environmental well-being and not driven solely by economic output. This mindset reflects a different kind of economics, one in which people matter the most. It takes inspiration from the work of alternative economists like E. F. Schumacher,[9] author of the now classic 1973 *Small Is Beautiful: Economics as if People Mattered*. Here's a brief glimpse into the issue of people in economics:

> *If we talk of promoting development, what have we in mind – goods or people? If it is people – which particular people? Who are they? Where are they? Concerns with people raise countless questions like these. Goods, on the other hand, do not raise so many questions. Particularly when econometricians and statisticians deal with them, goods even cease to be anything identifiable, and become GNP, imports, exports, savings, investment, infrastructure, or what not. Impressive models can be built out of these abstractions, and it is a rarity for them to leave any room for actual people. Of course, "populations" may figure into them, but as nothing more than a mere quantity to be used as a divisor after the dividend, i.e. the quantity of available goods, has been determined. The model then shows that "development," that is, the growth of the dividend, is held back and frustrated if the divisor grows as well. It is much easier to deal with goods than with people – if only because goods have no mind of their own and raise no problem of communication.*
>
> (E. F. Schumacher[10])

See the underlying issue? Most current economic analyses and approaches don't always address the human element. Again, not an easy thing to do, but required if communities are to build capacity and resiliency for a durable economy. Shifting the focus to people and the natural environment enables new ways of thinking, and new approaches that work better for both.

Witness the *new economics*[11] type approaches emerging globally – the Local First movement in the U.S.; Slow Food and Slow City movements in Europe; fair trade and sustainable seafood movements globally; socially responsible enterprises and even renewed interest in cooperatives and employee-owned business enterprises. There's much interest in all these areas with exciting works emerging from a variety of perspectives. For example, Bill McKibben's[12] *Deep Economy*, and *Eaarth: Making a Life on a*

Tough New Planet, and other provocative books in ecology and economics provide reason to pause and rethink our current development approaches. Or Frances Moore Lappe's[13] call to action in *EcoMind: Changing the Way We Think, to Create the World We Want* and the stunning, often shocking, revisit of food systems and development in *Hope's Edge: The Next Diet for a Small Planet*. Michael Shuman's[14] revelatory explorations into locally-focused economics in works such as *Local Dollars, Local Sense*, or *The Small-Mart Revolution* is helping spur a movement headed by the Business Alliance for Local Living Economies (BALLE). Essentially, proponents of new economics aim to "improve quality of life by promoting innovative solutions that challenge mainstream thinking on economic, environment, and social issues ... by working in partnership and putting people and the planet first."[15]

It's clear that the traditional methods of economic development popular since World War II don't always work best, and in many cases, not even enough to sustain communities marginally. Obviously, there are those places where industrial recruitment and other traditional approaches help economies thrive. But there are many communities that suffer when they lose their businesses and are at the mercy of external economic forces dictating their future. Does it always have to be so? Perhaps not, and the *new economics* way of doing economic development can work, fostering a resilient, sustainable and durable community. It takes considerable effort and continued commitment from all involved, a long-term focus, and attention to many elements of community.

> *Burlington is a place where change happens. That is why we are here.*
>
> Seventh Generation

About the place

Burlington embraces civic and social entrepreneurship. It's had a relatively stable population base for nearly 40 years; now with just over 42,000 residents in the city proper (the greater metro area has just over 200,000). But a stable population isn't enough to account for this resilient and durable economy. It has one of the lowest unemployment rates in the nation, (consistently through the years), the second lowest city foreclosure rate in the U.S., and property values holding when most of the country lost value during the Great Recession. How did this happen? Many communities have capacity building, leadership and economic development initiatives, so what makes Burlington different? What did it take? What worked and what did not? What are the key ingredients, processes, and strategies needed to successfully achieve community resiliency and build a durable economy?

An article by Kiplinger's Personal Finance Magazine in 2010 selected Burlington as one of the top ten cities for the next decade. Selected as #8, it ranks right up there with the big cities! Reasons cited for its selection include its green industry with companies like Seventh Generation, producers of earth friendly household products and clean energy production. Amenities like the Church Street Marketplace, a very successful pedestrian shopping district in the historic downtown core, and a lakeside location add to its ability to grow companies and jobs over the long term.

For more details and to see the full story, search for "The economy in our number-eight pick for Best Cities for the Next Decade is powered by the green movement," by Stacy Rapacon, Channel Editor, *Kiplinger's Personal Finance Magazine*, June 10, 2010.

A video link about Burlington is available at: www.kiplinger.com/fronts/archive/videos/ (type in "Burlington" to link).

Taking the long view

Political capability is a vital component of a durable economy. Having the ability to help guide a community and envision its future is essential; strong political capability can help provide this guidance. If civic and social entrepreneurship is a major foundation contributing to resiliency and durability, then leadership is the key ingredient. Burlington has long focused on civic and social leadership development and the results have been evident over the years. It started nearly 30 years ago, with the election of Bernie Sanders, now a Vermont U.S. Senator, as Mayor of Burlington. Bringing a fresh perspective, this was the turning, or "tipping" point. Ahead of the curve, there were points of focus such as an emphasis on sustainability which continue today that have shaped and molded Burlington into an award winning city. See the Appendices for some of these milestones and awards spanning nearly 30 years that help provide insight into the history of development in Burlington. It's about how to make Burlington a healthy place to live – economically, socially, and environmentally. And healthy it is: the Center for Disease Control (CDC) named it America's healthiest city in 2008 (see sidebar).

The Center for Disease Control names Burlington, Vermont as Healthiest City in the U.S.

By Mike Stobbe, AP medical writer

A vast majority of people who live in Burlington feel they are in good or great health! At 92%, this places Burlington as #1 in the nation, according to a survey by the Center for Disease Control. Combined with measures among the best in rates of exercise and low levels of obesity and diabetes, Burlington provides a healthy environment for its denizens (and visitors).

(For more details and to search for the full story by Mike Stobbe, AP Medical Writer, see *USA Today's* website at: www.usatoday.com/news/health/2008-11-16-burlington-health_N.htm)

Leading and leadership

People have been the focus of Burlington's efforts, and continue to be now. When Mayor Bernie Sanders established the Community and Economic Development Office (CEDO) in 1983, Burlington shifted its focus to developing leadership capacities in both the private and civic sectors. CEDO's mission has always been to foster economic vitality; preserve and enhance neighborhoods, quality of

Figure 1.2 Burlington Boathouse on the shore of Lake Champlain in Burlington, Vermont, which the CDC has named America's healthiest city
Photo: Nick Warner.

life and the environment; and promote equity and opportunity for all of Burlington's residents. By focusing on asset development, leaders exemplify civic entrepreneurship, developing collaborative and resilient relationships across the community and providing a platform for continued progress towards mutually desirable goals. One of Sanders' primary goals was to reinvigorate city government by creating more capacity for enacting an innovative economic agenda. This agenda combined supporting business with providing social assistance for underserved community members. From the beginning, it incorporated a *fourth sector* mindset, looking at both public and private business and enterprise as a way to help foster higher quality of life and positive outcomes. The fourth sector mindset can be thought of as a socially responsible way of doing business, or social enterprises with shared boundaries between the public, private, and nonprofit sectors. These businesses or enterprises are usually not driven predominately by a profit motive as they also have a socially purposed mission as well.

It isn't easy to build this type of leadership capacity, especially given changes in politics and politicians with election cycles. So what makes these efforts go beyond election cycles and have long-term endurance? It starts with asking a simple question of the residents: "What do Burlington residents want?" Keeping the focus on this generates longevity of intent. CEDO looked for innovative strategies and creative funding mechanisms to achieve its mission and guidance on how to do this. Identifying economic development priorities and strategies began with the *Jobs & People*[16] planning document in 1984. Continuous assessment and planning efforts are essential. *Jobs & People* has been updated four times through the decades, most recently in 2010. It provides the framework for identifying and carrying out economic and community development goals in the city – the blueprint. It has community engagement as a vital part of constructing the document and plans, going back to the essential question, "What do Burlington residents want?" It's all about engaging the past in a conversation with the present over a mutual concern for the future.[17] This conversation can be maximized, and continued.

Setting the sight: principles and goals

A variety of planning and coordination activities keep the focus in sight. A set of eight *economic development policy principles*, and the goals derived from them, inform the directions and dimensions of day-to-day project work. The policy principles are:

1. *Nurture sustainable development to provide for the city and its residents over the long term.*
2. *Promote and strengthen a mixed economy, and work actively to retain existing businesses and jobs.*
3. *Promote and support locally-owned and controlled small businesses including home occupations appropriate to the character of the neighborhood.*

4 *Partner with the private, nonprofit, and other government sectors to support existing businesses, attract future development, and conduct joint marketing.*
5 *Invest in the necessary public improvements, particularly transportation, to strengthen the downtown, both as a Regional Growth Center, and as a city neighborhood.*
6 *Work with neighboring communities, regional agencies, and state government to promote land use and development policies that support Burlington's role as the Regional Growth Center.*
7 *Support sustainable development activities in target areas of the city including the enterprise community, neighborhood activity centers, the Pine Street corridor, downtown, and the downtown waterfront.*
8 *Focus technical assistance, marketing and recruitment for economic development towards target industries.*

Next, CEDO calibrates these policy principles with various goals. These guide the development of individual strategies and form a solid framework for the recommendations outlined in *Jobs & People IV: Towards a Sustainable Economy (2010)*. The goals include (a) enhancing quality of life by maintaining a vital and diverse downtown and waterfront – a focus for the city's residents and visitors. The Church Street Marketplace is one of the most successful pedestrian malls in the U.S., its long-term success attributable not to the brickwork and physical amenities, but to a sustained community collaboration, with daily operations funded by a special tax assessment and with a full-fledged city department playing the lead role in maintenance, marketing, and promotion. Other goals are: (b) keeping existing locally-owned enterprises strong and supporting new enterprises that offer essential goods and services readily available to all residents; (c) enhancing Burlington's 350 acre agricultural breadbasket – the Intervale, home to market farming, and where community-supported agriculture, community gardens, and farmer training thrives; (d) sites with real or perceived contamination issues are redeveloped into productive use; (e) quality employment supports and opportunities are available for those who are traditionally underserved, and workers are earning a livable wage; (f) transportation needs are addressed, traffic congestion reduced, access in and around downtown improved and greater use of alternative modes of transportation promoted; (g) Burlington's competitive advantages are maximized by supporting the development of targeted industries, including tourism, telecommunications-intensive businesses and the environmental technology industry, financial services, specialty foods, media, the arts, and sustainable natural resource promotion; and (h) new cooperative relationships are developed between the City and other communities in the region to strengthen the regional economy for the benefit of all.

Envisioning the future

Looking at this collection of goals, it is clear where progress has been made, and where more is needed. Burlington isn't perfect, much remains to be accomplished and remedied. Pervasive problems like poverty among single mothers in the city or rising costs of living outpacing income levels are just two examples that need continued attention. But on balance Burlington has done well. The strategy of small-scale, indigenous-type development has been successful. Because the population has remained relatively stable, issues related to growth have not been overwhelming. Progress is gauged via ongoing evaluation efforts, and through a long-term planning effort for sustainable development. Beginning in the 1980s, sustainable community development principles were added to the plans (see Closer Look section for a listing).

In 1999, a long-term vision was added to the mix, one that could be embraced by the entire community. The Burlington Legacy Project brings all sectors of the community together for its 2030 vision. Adopted in 2000 by the City Council, its Legacy Action Plan[18] defines key priorities, actions, and indicators. The Plan, reflecting input from hundreds of community members, is based on five major themes: Economy, Neighborhoods, Governance, Youth and Life Skills, and Environment. Its vision is to:

- Maintain Burlington as a regional population, government, cultural, and economic center with livable-wage jobs, full employment, social supports, and housing that matches job growth and family incomes.
- Improve quality of life in neighborhoods.
- Increase participation in community decision making.
- Provide youth with high-quality education and social supports, and lifelong learning opportunities for all.
- Preserve environmental health.

Frequent updates gauge progress towards goals. See for example the "Legacy Report Card" on the City of Burlington's website. Staying in touch with goals and progress is essential, as many changes can occur within a community. The next section addresses some of the stresses in a system and the importance of maintaining leadership and political capability throughout times of challenge.

Keeping space: sprawl, politics and sheer willpower

Burlington isn't immune to issues of sprawl and patterns of development that endanger its sense of place, character, and charming attributes. In the 1990s, an adjacent municipality decided to allow a large mall development that would have adversely impacted Burlington's local businesses as well as contribute further

Figure 1.3 Aerial view of Burlington looking south along Lake Champlain
Photo: Kirsten Merriman Shapiro.

to sprawling patterns of development in the outlying areas. An epic battle ensued, one in which the political leadership, as well as the sheer willpower of residents and local leaders, prevailed. Seminal research was created by the city in response to the legal battle; this included insights on:

1. *The impact on public investment of sprawl, estimated at $100 million of public investment at risk.*
2. *The impact on local tax rolls since retail pays more per square foot in property taxes.*
3. *Traffic impacts, and who pays for needed public improvements such as roads, bridge replacements, police and fire stations.*
4. *The loss of jobs as a result of big box store development, low wages and lack of good benefits (for example, one of the big box companies has many of its employees on welfare since wages are not high enough and benefits are limited).*
5. *Predatory pricing investigated by the Congressional Research Service at Vermont's Senator Leahy's request –*

found instances where big box retailers drove out family owned and small businesses via predatory pricing practices.

Community education efforts were developed including ad campaigns; music videos; production and distribution of T-shirts; development of a membership organization in the adjacent municipality; rallies; investigative documentaries on the impacts of sprawl; and so on. More than 200 articles were printed in the local newspaper looking at myriad perspectives. Final results? Some of the development was still built, but with 50,000-square-foot limits on size. After winning this battle, the City was asked by dozens of other communities to help them fight big box retail. Another outcome? The Vermont Forum on Sprawl was founded as a result, and their first director was one of the consultants who had conducted some of the primary research.

Taking all these considerations discussed in this chapter, what's the bottom line, so to speak? It really comes down to: *process*, *relationships*, and *results*. In other words, the underlying principles guide efforts to create good *process*, helping forge strong collaborative *relationships* that ultimately produce *results* in the community. Having responsive political capability and leadership helps produce desirable results and outcomes.

A closer look: Burlington's journey

Former mayor Peter Clavelle gave the following speech in 1995, and it still holds much relevance for community economic development. Many of the suggestions and areas of focus in the speech have been implemented in Burlington, so the ideas presented can be related to outcomes realized in the community. The presentation discusses long-standing sustainability principles, and offers suggestions to other communities wanting to enhance resiliency and durability. The speech touches on many key aspects for fostering a strong political economy sector that, in turn, helps foster a more durable and resilient economy. Most importantly, this speech represents an expression of what Burlington residents wanted for their community. Given the continued support for sustainability principles, it also seems to represent what they still want.

Development and local government

Economic development, and the role of local governments in stimulating economic growth, is being redefined around the world. Two factors are principally responsible for changing the direction of economic development at the local level:

1. The resources available to local governments are shrinking. Government alone cannot solve the problems facing urban communities. Increasingly, government is adopting the role of catalyst and facilitator. This new role necessitates the interweaving of scarce public and private resources and the creation of partnerships with the private sector and the social sector. This role also requires that the citizens of a community be empowered.
2. The interconnectedness of economic growth and environmental protection is becoming increasingly obvious. More and more communities are embracing strategies of sustainable development in recognition of this reality. While numerous definitions of sustainable development exist, the most popular is "to ensure that development meets the needs of the present without compromising the ability of future generations to meet their own needs."

Before sharing with you my views and experience on this issue, I'd like to tell you who I am and where I'm coming from. For two terms, I served as Mayor of Burlington, Vermont – a city in the northeast United States. Burlington city government has a national reputation for citizen involvement and innovative ideas. We are also a unique city politically. In a country dominated by two political parties, Burlington, Vermont, for the last thirteen years, has had as its leading political force the Progressive Coalition - an independent third party. The Progressive Coalition was born in the early 1980s in an effort to open City Hall to everyone. They have been able to dramatically improve the responsiveness and efficiency of city government. The achievement of equitable and sustainable community development has been an overriding goal of the Coalition's agenda.

Prior to being elected mayor, I served as the City's Community and Economic Development Director for six years. Most recently, I have resided in Grenada where I was involved in community development and political consulting. I've also worked in the Gaza Strip to strengthen the capacity of local governments there.

Cities and states, in the U.S. and around the world, have seen their social and economic woes grow worse during the 1980s and early 1990s. At the same time, for U.S. cities, many of the federal programs that might have alleviated those problems grew scarce. As Washington drew back from economic planning, human services, affordable housing and many other community development concerns, local governments were expected to take up the slack.

Most cities and states did what they could. Forced to do more with less, however, many soon found they could do very little to arrest community decline or to promote community development. Some stopped trying to do community development at all.

The story was different in Burlington, Vermont. An activist municipal government, working in partnership with the private sector and a network of municipally supported nonprofit organizations, pursued a sustainable development strategy before the term was invented. The strategy involved generating new sources of public revenue; creating and retaining jobs; encouraging and regulating growth; ensuring a publicly controlled waterfront; producing permanently affordable housing; stabilizing residential neighborhoods; reducing energy consumption; promoting – in fact requiring – the recycling of solid waste; and removing barriers preventing women from enjoying the fruits of economic growth.

In Burlington, *sustainable community development* is guided by six public policies or principles. They are:

1. Encouraging economic self-sufficiency through local ownership and the maximum use of local resources ("seal the leaky bucket");
2. Equalizing the benefits and burdens of growth;
3. Leveraging and recycling scarce public funds;
4. Protecting and preserving fragile environmental resources;
5. Ensuring full participation by populations normally excluded from the political and economic mainstream; and
6. Nurturing a robust "third sector" of private, nonprofit organizations capable of working in concert with government to deliver essential goods and services.

I suggest these principles make sense whether you live in Burlington, Vermont, Fortaleza, Brazil, or St. George's, Grenada.

While all principles are critically important, I feel particularly strong about the need to create and nurture a strong social sector. Comprised of nongovernmental organizations or institutions necessary for sustainable development, this sector assures empowerment and participation which are prerequisites for sustainable development. This sector is different from state capitalism or state socialism. As described by Osborne and Gaebler (1992, p. 44):

> This sector, it seems to us, is made up of organizations that are privately owned and controlled, but that exist to meet public or social needs, not to accommodate private wealth.

Please do not hear me to suggest that the role of the more traditional for-profit, private sector is unimportant or minimal. I emphasize that the social sector, as this growing network of nongovernmental, nonprofit, and voluntary organizations, is playing an increasingly important role in community and economic development. The lines between the three sectors will become increasingly blurred.

Let's get practical. What can local government do – in cooperation with the private sector and the social sector – to bring about desired economic development outcomes?

First, you need a strategic economic plan – a plan that will serve as a blueprint or road map in designing and implementing economic development programs and actions to help your community effectively meet future challenges and enhance its capacity to achieve shared and sustainable growth. The plan needs to understand SWOT – strengths, weaknesses, opportunities and threats. The plan must address the challenges facing a community: ensuring that residents left behind are given the resources and support necessary; ensuring economic well-being is sustainable by planting the seeds for the continued generation of quality jobs through self-employment, enterprise development, and small business growth – working towards the creation of a locally-owned and controlled economy; maintaining the vibrancy of downtowns and central business districts and discouraging suburban sprawl; creating cooperative relationships within a regional economy that recognize the interdependence of a metropolitan area with the central city; building on a city's competitive advantages by supporting the development of targeted industries – telecommunications-related, arts-related, sports, tourism, and environmental industries.

The plan must be clear about strengths and weaknesses. What is the telecommunications infrastructure, access to suppliers, utilities capacity, availability of bank loans, quality of life, availability of labor, energy supply, availability of land and buildings, quality of schools, access to transportation links, tax structure, and ability to attract skilled employees?

What are the needs of the residents of your community – the labor force? What barriers are faced by the unemployed and underemployed of your community? In Burlington the trends were: increasing rates of poverty (particularly among single mothers), suburbanization of not only the affluent population but also many employment opportunities and disappearance of many traditional industries. Specific barriers faced by our residents included child care, worker retraining, and transportation.

I'd like to briefly review specific economic development strategies. While some may not be relevant to your community, I believe many are.

1 *Ensure the continued growth and development of locally-owned businesses*. Local authorities can play a very significant role in the start-up, expansion and retention of locally-owned small enterprises. Support for micro-enterprise development and the nurturing of

entrepreneurial capacity should be the cornerstone of a community development strategy. I'll further discuss this strategy in a moment.

2 *Build on a community's competitive advantages by supporting targeted industries.* This strategy will vary from community to community. What's important to understand is that every community should be strategically targeting economic activities that fit the competitive advantages of a location. It's important that these "sectoral" targets conform to the residents' vision of their community and respond to both real market opportunities and real constraints. We are shifting from large-scale manufacturing to a knowledge-based sector. Other opportunities include: food processing, financial services, cultural and entertainment sectors.

3 *Improve economic well-being of targeted population groups.*
   - What are underserved population groups (women, single moms, minorities, working class youth)?
   - Strengthen women's economic opportunity programs (STEP-UP, Women's Small Business Program [see Chapter 3 for more details]).
   - Expand youth employment programs for non-college-bound kids (job development and basic skills).
   - "One stop" service centers for job training and adult education.
   - Upgrade transportation service for low-income residents; expand pre school and daycare.

4 *Maintain strong and vital downtown.*
   - Downtown should be a center of regional commerce, professional activity, and recreation and culture.

- Block development of suburban mega retailers and prepare downtown stores for substantial new competition. Two-prong – (1) offensive; (2) defensive – enhance cultural, recreational and entertainment amenities. Examine and adjust marketing, merchandising, product offering.
- Address downtown transportation needs. Multi-modal, connections, parking.
- Strengthen institutional capacity to maintain and enhance downtown retail.
- Enhance job opportunities in retail sector for residents. Work with employment and training providers: (1) recruit hard-to-employ low-income residents; and (2) offer "move-up" and "move-out" opportunities.
- Provide infrastructure: sufficient transportation access and utilities.

5 *Continue to invest in economic infrastructure.*
   - Transportation access;
   - Informational signage;
   - Improve traffic flows (increase road capacity and improve traffic management);
   - Upgrade water and sewer;
   - Electrical/solid waste.

6 *Develop new cooperative relationships to support local and regional economic development efforts.*
   - Establish a community development corporation. Centralize initiatives in business development, job training, and even housing development. Advantages: administrative cost-effectiveness, "synergy," and less politicized environment.
   - Promote increased regional cooperation – increasing interdependence on central city economies and surrounding communities. Task

Force – airport, regional approach to new enterprises, business financing.

7 *Growth and development of small business.*

- Small business is where the action is. Jobs are continuing to be lost by the Fortune 500. Economic growth is being driven by small business. Increasing emphasis on micro-enterprises – businesses with fewer than five employees that require only small amounts of capital. Preference for local ownership – key actors are residents. More likely to hire locals, use local resources, and committed to community. Specific preference for employee-owned. Empowerment. Share in fruits of labor. Small business development addresses three agendas:
  (a) Job creation – maintain economic growth often by offsetting effects of recession.
  (b) Community rebuilding – creation of vital economies: persistent poverty or economic dislocation – structural change.
  (c) Economic self-sufficiency – improve the ability of individuals to participate in economy. Personal development strategies to aid this as well (financial management, communication skills, etc.).

8 *Program models – micro-enterprise development.*

(a) Entrepreneurial training and support – focuses on training business owners.
(b) Comprehensive business development – personal development, business planning, and lending.
(c) Incubators – commercial space leased at below market rates with shared services.

(d) Lending – provides credit to potential entrepreneurs. Peer lending or individual.

(e) Market research – attempts to improve odds of success for start-up. Target based on consumer demand/niche.

(f) Cooperative or sectoral business development. Employs collective approach. Selects industry with potential to employ large numbers within cooperatively owned enterprise.

9 *Financial aspects.*

- Financial and technical assistance: capital is essential ingredient in every phase. Revolving loans – fill "gaps" in private market. Seed money is needed too.
- Targeted to specific neighborhoods with levels of economic stress. Priority to firms employing low and moderate income workers.
- Micro-enterprise program – capital and technical assistance; peer lending with peer pressure; support and self-monitoring; loan decisions are made by group – group is collectively responsible; private sector debt financing and private risk capital sources are essential.
- Linked City deposits – City funds deposited in local bank; bank commits to community reinvestment activities.
- Community banking council – bankers, the City, businesses, and nonprofits identify and respond to unmet credit needs.
- Establish a risk capital fund – small equity-like investments. Source – local pension funds, social investment, private contributions, bank funds.
- Encourage home-based business.

- Fostering continued development – growing role with advances in computer and communication technologies; newsletter (networking and access to resources), business curriculum, access to office equipment and services, mentoring.

10 *Increase competitiveness.*

- Joint marketing network for business service firms;
- Regional consultants network;
- Institutionalize "calling" program to visit with businesses and identify needs and opportunities; and
- One-stop city permitting – time consuming, costly, frustrating.

There you have it! I want to close by emphasizing a few points: government alone cannot be responsible for stimulating economic development as an era of partnerships is here. Government's role is as catalyst and a facilitator – empowering rather than serving should be its goal.

(Appendix B outlines 200 initiatives that were accomplished over a 30-year period, most of which are outlined in this speech.)

Resources and ideas for making it happen in your community

http://burlingtonvt.gov

Master/comprehensive community planning: Read the master plan for your community. When the plan is being updated recommend changes to protect and enhance community assets. Be aware there is an inherent conflict between housing and industrial uses. Maintain an area for expansion of growing businesses.

Long-term development planning: Create a long-term strategic plan based on market research, existing data, and personal interviews. Look to build on assets and reduce outflow of capital from your community. Present plan and outcomes to the community at least every other year.

Organizational context: Understand how your city is organized and what it is authorized to do and who has the authority over different functions. Review your city charter and any other documents that will help you understand how things work.

Current events: Read legal notices and want ads frequently to stay on top of projects underway and find out who is hiring.

Professional development: Read professional journals and magazines to stay on top of trends.

Media relations: Nurture relationships with the media. Go meet with your local daily newspaper publisher and editors; open up lines of communication.

Build credibility: Choose a project that you can execute successfully to build credibility in year one. Do the same in year two. Slowly develop more complex projects over the years.

Develop directories: Develop directories of all the organizations providing services in your community, including workforce training programs, community lenders, business technical assistance providers, business support programs. Distribute them at public places such as the polls on election day and post on them on your website. The directories will multiply your effectiveness.

Develop new programs: Map out the local and state resources available to support business development and compare to other states in your region. Meet with key legislative leaders and review assessment. Develop a new program modeled on best practices in a neighboring state and pass a law to support that effort.

Electing officials: Help an elected official get re-elected, or run for elected office. Engage in the political process.

Work with local college alumni office: Encourage local alumni from schools to establish businesses. Students who go to local colleges and universities have an affinity to your community and might want to start a company. They also might not be able to find a job, so starting a business might be a viable, attractive option.

Records: Keep detailed records of projects and all contacts you have made during your career. Over the years your effectiveness compounds with this invaluable database of contacts.

Stay in contact: Contact major businesses, bureaucratic and political leaders to see how they are doing and to find out how you might help them. Do this annually.

Invest in fundraising: Hire a person in your community to write grants. This investment usually more than pays for the expense.

2 Localize and Socialize

How Does Fostering Locally-focused and Socially Responsible Businesses Help Create a Durable Economy?

The term vernacular is used to "convey that the people who create the culture and the businesses must own the culture and be rooted in place."[1] At the same time, communities evolve and change through the years, as people and events flow through them. This definition of vernacular as being rooted in place can be interpreted as being particularly supportive of locally-owned businesses as opposed to absentee owners or remotely controlled enterprises (large big box or chain stores, for example). This definition also implies that both authenticity and historic culture can serve as the basis for community economic development approaches. Carr and Servon[2] (2009, p. 30) explain this ability as:

> *An economic development strategy grounded in vernacular culture achieves a balance between small, cultural diverse businesses and larger chain establishments, develops and celebrates the historical character and nature of the community to make it attractive to residents and investors, and fosters uniqueness that serves the needs of the resident community and likely attracts outside shoppers and tourists.*

Underlying this premise is the belief that a community's long-term success hinges on this idea of creating and owning its culture. How better to do that than to focus on locally-owned businesses, socially responsible and vested organizations and a preference for governance versus government? This chapter explores the concepts and applications of the "power of local," and the notion of creating a business culture that survives short-term economic and political changes to endure through time. The first section centers on locally-focused enterprises and the second explores socially responsible businesses (the "socialize" part of the chapter's title). The Closer Look selection describes Burlington's Local Ownership Development Project. A locally-focused economy sector brings together many of these dimensions including socially responsible businesses and culturally competent businesses and organizations.

Further, it is important that the people in the community have a voice in the process of development. Part of the premise of local ownership is that people will support and engage in their communities, becoming involved with decision making and otherwise helping guide the forces shaping the area. Governance structure and organizations should provide a vehicle for people to express their needs and desires, and showcase what they do. Concerns are then identified and elected officials must respond when enough voices are heard. Local ownership and citizens vested in a locally-focused economy sector are better positioned to positively influence the direction and course of policy outcomes. In addition, local ownership and a locally-focused economy component fosters a fourth sector environment where social purposes can integrate with businesses and vice versa.

Why local?

Michael Shuman, cofounder of the Business Alliance for Local Living Economies, explains in *The Small-Mart Revolution* that local ownership is an essential condition for community prosperity for at least five reasons:[3]

1 Locally-focused businesses are long-term wealth generators. Many entrepreneurs are in a particular community because they love living there and this makes them less likely to leave. The longevity of some of these businesses can span several generations.

2 *Fewer destructive exits.* Massive upheavals can occur in a local economy when a large employer exits, creating a "death spiral" where a sudden exit is followed by high levels of unemployment, shrinking property values, lower tax collections, and deep cuts in schools, police and other services. Economies comprised of locally-owned businesses are far less likely to experience this dramatic decline.
3 *Higher labor and environmental standards.* Local quality of life is better protected in communities made up mostly of locally-owned businesses, via shaping of its laws, regulations, and business incentives. Locally-owned businesses do not typically threaten to leave town, and can set reasonable labor and environmental standards with confidence. Further, business incentives can be tailored to the needs of the majority of locally-owned businesses, rather than giving large subsidies to nonlocal businesses. It is Burlington's experience that locally-focused businesses are more responsive and more vested in the community.
4 *Better chances of success.* Locally-owned businesses are not as susceptible as large companies to a temptation to move when costs rise – witness the lure of offshore manufacturing locations for many industries throughout the U.S. Moving or relocating to another community or country simply isn't an option for many locally-owned businesses.
5 *Higher economic multipliers.* Studies show that the impact of a dollar spent at local businesses has a far greater impact than money spent at chains or big box retail. Local businesses yield two to four times the multiplier benefit as compared to nonlocal businesses. Local businesses have higher multipliers because they spend more locally. In other words, local management uses local services, advertises locally, and enjoys profits locally. To illustrate, there is only one franchise in the National Football League owned by a

community-controlled nonprofit with shareholder members (primarily residents of Wisconsin). While other franchises can and do leave their host communities, the Green Bay Packers are a critical source of wealth and economic multipliers for Green Bay and will be around for the long term.

Here's another reason too: local economic ownership can improve local prosperity because these types of enterprises support the transition to a more sustainable economy (both local and global). When they're supported by government policy to focus on producing more needed goods and services that otherwise might be imported at a higher total cost (especially food, energy, and affordable housing), the economy becomes stronger and more resilient. Burlington has been at the forefront of developing policies and programs providing things people need locally rather than importing them (energy and food to name just two).

Further, a local focus tends to start off incrementally, with small businesses. The ability to generate, support, and grow local businesses is key to building successful economies. It is worth reiterating: small business growth – particularly start-ups – is what drives economies not only in Burlington but just about everywhere else too. The Kauffman Foundation recently released a report focusing on just that (see sidebar for excerpt). They found that, "because start-ups that develop organically are the principal driver of job growth in the economy, job-creation policies aimed at luring larger, established employers inevitably will fail." Focusing on creating a nurturing environment where start-ups and small businesses can thrive is vital. It's no longer enough to offer cursory services to small businesses, it takes concerted action and deliberate intention to provide the level of service historically accorded to recruiting existing businesses throughout the U.S.

Job growth in U.S. driven entirely by startups, according to Kauffman Foundation study

Ewing Marion Kauffman Foundation press release 7/07/10. http://bit.ly/byg7Kh

Although conventional wisdom suggests that the annual net job gain at existing companies is positive, the fact is that net job growth in the U.S. economy occurs only through start-up firms, a new report from the Ewing Marion Kauffman Foundation (www.kauffman.org/) finds. Based on the U.S. Census Bureau's business dynamics statistics, the report, *The Importance of Startups in Job Creation and Job Destruction*, found that both on average and for all but seven years between 1977 and 2005, existing firms were net job destroyers, losing a combined one million jobs per year. In contrast, during their first year new firms added an average of three million jobs. The report also found that while job growth patterns at both start-ups and existing firms were pro-cyclical, there was much more variance in job growth patterns at existing firms. Indeed, during recessionary years job creation at start-ups remained relatively stable, while net job losses at existing firms were highly sensitive to the business cycle. And it's not just net job creation that start-ups dominate. Although older firms lose more jobs than they create, the gross flows decline as firms age. On average, one-year-old firms create nearly one million jobs, while ten-year-old firms generate only 300,000. In other words, the notion that firms bulk up as they age is not supported by data. Because start-ups that develop organically are the principal driver of job growth in the economy, job-creation policies aimed at luring larger, established employers inevitably will fail, said the report's author, Tim Kane. Such city and state policies are doomed not only because they are zero-sum but because they are based on unrealistic employment growth models, added Kane. "These findings imply that America should be thinking differently about the standard employment policy paradigm," said Robert E. Litan, Kauffman Foundation vice president of research and policy. Policy makers tend to focus on changes in the national or state unemployment rate, or on layoffs by existing companies. But the data from this report suggest that growth would be best boosted by supporting start-up firms.

The local scene

With just a bit of investigation, it becomes clear why local is preferable to other types of businesses that may not be as vested in a community. Burlington got lucky in this respect with an influx of young progressive thinkers and doers in the 1960s and 1970s. Many of them started businesses out of necessity to create jobs for themselves, and because Burlington provided a supportive environment in which to try new business ideas and concepts. These residents are committed to Burlington and to Vermont – fully vested in their communities. Over time, most have provided more than business or economic benefits to the area, helping establish educational, social or environmental charities or outreach initiatives. "Burlington has always felt like a land of opportunity since day one," says Steve Conant, who moved to the area as a student and has been a long-term small business owner and social entrepreneur. When employment prospects were slim, he established a business in one of the Pine Street incubators. Thirty years later, his Conant Metal and Light Workshop is a fixture on Pine Street, where he has played a leadership role in the South End Arts + Business Association (see the Closer Look case in Chapter 4).

One of the seven key recommendations of the 2010 *Jobs & People IV* (the long-range development plan for Burlington) is for CEDO to continue to focus on Local First (note the recommendations in this plan call for a long-term focus of 20–30 years). This is not only about businesses producing goods and services locally and using locally-sourced materials where possible, but also changing buying habits of people so that they consume these locally-produced goods and services. This latter part of changing consumer habits needs to be promoted through various marketing initiatives such as the proposed new "Fresh District" highlighting Burlington's commitment to urban agriculture production (including Intervale, see Chapter 7) and value-added food production.

CEDO's history of support for locally-owned and cooperative enterprises dates back to 1984. Working directly with businesses that produce goods and services locally, or from locally-sourced materials, is a primary focus. A second step in this strategy is the promotion of resident, nonprofit, and business purchasing from local sources. CEDO's projects relating to Local First include:

- Intervale Center
- Energy Co-op of Vermont
- Fresh District

- ONE Coupon Book and Guide
- Business Outreach
- Food Enterprise Center
- Value-added Food Sector
- ONE World Market
- Legacy Plan
- Burlington Bread (complementary currency).

(see www.cedoburlington.org for more info and links to these programs)

Incubating business

Private owners developed the first incubators in Burlington early on, in the 1970s. CEDO followed suit, using the Urban Development Action Grant (UDAG) program to renovate the Maltex Building into an incubator which could house a number of small businesses. CEDO continued to support the redevelopment of 392,601 square feet of vacant or underutilized buildings into incubator buildings in the area in the late 1980s, including the old Soda Plant at 266 Pine Street (using industrial revenue bonds), the Vermont Maid Maple Syrup building (using the Burlington Revolving Loan Program), Flynn Dog at 208 Flynn Avenue (business and technical advice), the Howard Space Center, and 294 N. Winooski Avenue, and the Kilburn and Gates building. These buildings have been instrumental in providing affordable space and support for hundreds of new and fledgling businesses, with conceptual, financial, and technical services.

Burlington is noted for its homegrown businesses, including Ben and Jerry's ice cream enterprise. "We don't think we could have achieved it elsewhere – because of its size and real ethic of support for local businesses here, it worked," Ben Cohen explained; "we moved here because no other ice cream parlor was here, we thought we'd maybe eke out a living but the cold season made it necessary to sell to others (beyond Burlington)."

> *I believe Burlington is resilient because our economic activity is small business, not a boom or bust situation since our sectors are diverse.*
>
> (Ben Cohen, co-founder of Ben and Jerry's)

It goes beyond serving as a nurturing environment for growing businesses; there is a distinctive essence about Burlington that transmits to its businesses, and vice versa. There are quite a few socially or environmentally focused businesses that further this essence, such as Seventh Generation, the natural cleaning products company. It's almost a brand, in some ways, alluding to the appeal of a green, clean (and a bit crunchy) environment. It's what some call "concept marketing," where developing a niche or specialty forms the basis of creating brand recognition for a community.[4] This isn't an exploitive approach – in this case, it's what has emerged quite naturally and organically in Burlington and in Vermont. As Ben Cohen explains, "There's an agricultural, family farm image in Vermont – this is a driver of economic development, and businesses are trading on that image." Here's a quote from a business in Burlington that reflects that value of the brand.

> *So many small companies, taking a grassroots and pragmatic approach to business, have leveraged Burlington's creative energy to build emotionally charged, socially purposeful, legendary brands. People from all over pay attention to the brand stories told from Burlington, so it made sense to me to tell the genuine story of Terry and grow the Terry brand from here.*
>
> (Liz Robert, Terry Bicycles)

"Young and vibrant, accepting and a great place to start a business," says Jim Lampman, creator of Lake Champlain Chocolates, when describing Burlington. His company is now one of Vermont's leading tourist attractions. The company now ships its premium chocolate products to many other areas. Locating here over 30 years ago, Jim says it is an area where "your voice can be heard, and that's important when trying to grow a business." Lake Champlain Chocolates is one of the success stories illustrating several concepts important for supporting flourishing business development. The first is providing small business support services and ensuring that such services are both needed and effective. Second, the business located its production facilities in the South End Arts + Business District (SEABA, see case in Chapter 4) which serves as an industrial and arts business incubator district. Third, the company ties in well with the branding and image of Burlington and Vermont, with a recognition now far beyond the borders of the state. Fourth, many refugees have been hired and are paid livable wages. This is an issue of concern in Burlington as a political refugee relocation city; for additional information see Chapter 3.

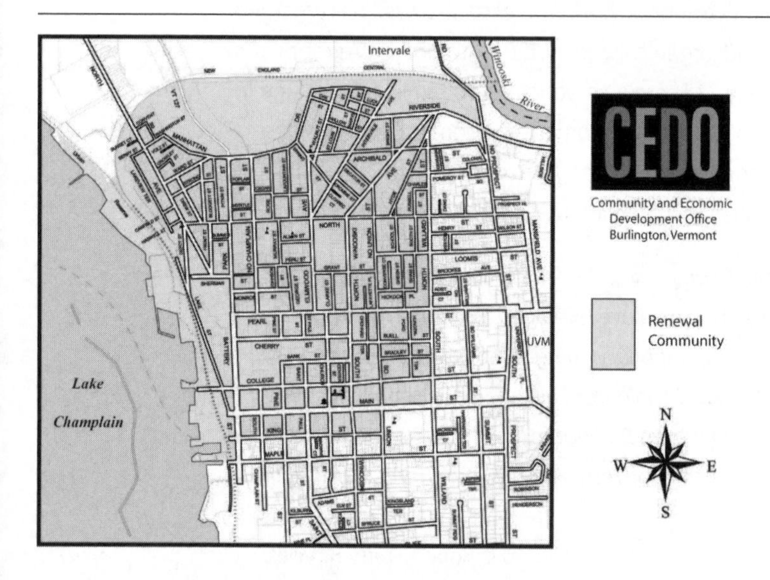

Figure 2.1 Map of Burlington showing the Renewal Community area

The Renewal Community designation, through the U.S. Department of Housing and Urban Development (HUD), provided federal tax incentives to spur economic development in the downtown and Old North End areas of Burlington. By 2009, almost $23 million in Commercial Revitalization Deductions (which allow for accelerated depreciation) has been awarded to assist with the construction/rehabilitation of over 198,000 square feet of commercial space. Telephone surveying among Renewal Community businesses suggests that the wage credit – up to $1,500 per year for each employee who lives and works in the Renewal Community – is one incentive that businesses are finding to be most useful. It is estimated that Burlington businesses have saved over $1,039,000, during the period 2002–2009, through the use of the wage credit. This is a conservative estimate.

Fifth, the company donates 10 percent of its pretax profits back to the community. Sixth, management gives its time to the community by serving on boards of local nonprofit organizations supporting the arts, business development, and programs for disadvantaged youth.

> *Burlington is a perfect fit with our brand. The traffic and visibility here helps our sales while the natural beauty and lively culture attract employees who are able to live close to where they work. Then, of course, there's Lake Champlain itself ... where would we be without the Lake?*
>
> (Jim Lampman, Lake Champlain Chocolates)

More of the good stuff: socially responsible businesses

Here's another reason too that a local focus is important: locally-generated and owned businesses tend to reinvest in their communities beyond the purview of their own business arena. Social, environmental, educational and other dimensions of community are important and are reflected in the businesses' activities and initiatives. There is a long history of social responsibility in the area.

Vermont was the first state in the nation to establish an organization of business owners for this purpose. Vermont Businesses for Social Responsibility (VBSR) works to strengthen the social and environmental infrastructure in Vermont by creating a strong climate for sustainable business growth. It was established with CEDO staff assistance and funding in 1990.

Vermont Businesses for Social Responsibility has members representing a variety of sectors across the state. All together, 1,200 members employ more than 14 percent of Vermont's workforce and generate more than $4 billion in revenue annually and 60 percent of members have been in business more than ten years.

(http://vbsr.org/index.php/pages/static/cat/about_vbsr/)

VBSR has the overriding mission of advancing business ethics that value multiple bottom lines on the economic, social, and environmental fronts. "We strive to help members set a high standard for protecting the natural, human and economic environments of the state's residents, while remaining profitable," is how VBSR describes it. They pursue their mission through three areas of activity:

> 1 *Education.* Empowering members to solve environmental, social and economic problems by providing concrete resources and information to help improve members' business practices.
> 2 *Public influence*. Initiating tangible change in public policy that combats exploitation and promotes sustainability by working to capture the inherent power of business to represent socially responsible ideals to legislative bodies, news media, and the general public.
> 3 *Workplace quality.* Enabling every worker to live and work with respect and dignity by creating work environments and economic climates that provide fair income in safe work settings, and allow each employee to contribute to a high quality product or service.
>
> (http://vbsr.org/index.php/pages/static/cat/about_vbsr/)

VBSR has become a strong influence, with a political arm to focus on state policy. This work includes encouraging passage of a range of state legislation for the Sustainable Jobs Fund, family leave legislation (which a year later was adopted by the U.S. Congress), Farm to Plate funding to encourage farm initiatives – there are 20–25 policy papers annotated on their website. VBSR is committed to being "never against anything and always for something." They provide an annual Legislative Scorecard, for members and others to see how legislators voted on each year's VBSR business agenda.

Here's a bit about the history – it's important to show the challenges and impetus for moving forward when forming advocacy and support organizations such as VBSR. It started by asking, "How can we make it work?" when a panel convened at Seventh Generation, the natural household cleaning products manufacturer. Participants included local business leaders such as Will Raap of Gardener's Supply and Jerry Greenfield of Ben and Jerry's. Bruce Seifer of CEDO challenged them and the other attendees to be actively involved in politics, encouraging them to tackle the political side of issues traditionally not pursued by the business community. At this meeting, it was decided to focus on political action. CEDO provided some of the start-up funding to hire a director for this new organization and later, in 1993, a public policy staff position was created as well. It's interesting to note that times and receptivity to corporate and social responsibility have changed quite a bit – when first trying to promote increased minimum wage to state legislators, the VBSR policy representative was literally thrown against a wall by a business lobbyist!

The focus on local has always been strong in Burlington and indeed, the entire state of Vermont. Growing out of VBSR is Local First Vermont, a nonprofit organization focused on preserving the character and prosperity of Vermont's economy, community networks and natural landscape. Local First Vermont's mission is to "preserve and enhance the economic, human and natural vitality of Vermont communities by promoting the importance of purchasing from locally-owned, independent businesses." The organization sponsors a variety of education programs and enhances marketing efforts, for example by posting Local First Vermont decals in merchants' storefronts or by offering special coupon books to spur interest in locally-focused businesses.

Local First

Local First initiatives across the U.S. are increasing, including efforts to promote locally-focused businesses with special marketing, such as the "Buy Local" coupon book. Here's how Local First Vermont describes its effort:

> The Local First Resource Guide and Coupon Book includes local coupons, business stories, and buy local facts. There are 141 contributing businesses from Addison, Chittenden, and Washington counties. 104 businesses will have coupons available in the book for a total of $2,300.00 in local savings. Not to mention the Burlington Old North End section where local fare is international and includes a Himalayan Market, Vietnamese Grocery, Middle Eastern Treats and more. All this, and the chance to support your local Vermont economy for 10 dollars.
>
> (http://vbsr.org/index.php/pages/static/cat/local_first_vermont/)

Another benefit of engaging in socially responsible business practices is that it tends to help set a positive tone for an area. Benefits accrue in an environment where social, natural, and cultural dimensions of community are important and recognized by the private sector. When socially focused or socially responsible businesses locate or generate from within an area, it serves to attract others of like mind to grow their businesses in the community. Here's a case in point:

> *We looked at various sites both in and outside of Vermont before selecting Burlington as the location for Burton. There were many reasons for the selection – the support provided by the City for small business, the available work force, the well-serviced airport, and the proximity to the highways.*
>
> (Jake Burton Carpenter, Burton Snowboards)

A dimension of socially responsible businesses trending now is alternative organizational structures for fostering a for-benefit orientation. Low-profit Limited Liability Company (L3C) and Benefit Corps (B Corps) are two such newer corporate structures. Vermont is the first state to enact legislation for L3Cs. The low-profit limited liability company is a hybrid between a nonprofit organization and a for-profit corporation, identified as low-profit with charitable or educational goals – any business in any state can register in Vermont. Also, Vermont was the second state in the nation to pass legislation for establishing benefit corporations. Benefit corporations are directly in line with a social entrepreneurship orientation and look to balance the company's shareholders' interests with the community's stakeholders' interests. According to *Bloomberg News*,

> *Having official "benefit corporation" status allows entrepreneurs to consider stakeholders like employees, communities, or the environment in business decisions. Under existing corporate law, company directors can face lawsuits if considering outside stakeholders is seen to damage the financial interest of shareholders.*[5]

Owning a community's future

Local business and industry need support, encouragement, incentives, and control.

> *Fortunately, despite whatever encroachments have been made by dispassionate big business,*[6] *we still have our local economies. We don't need to take them back from global corporations; we already have them, in whatever condition they're in – good, fair, or poor. We can move them forward from here. And if we can keep local institutions and businesses alive ... we will keep the continuity of generations alive and maintain the richness of our communities.*[7]

John Abrams discusses the importance of locally-focused businesses in his book, *Companies We Keep* – it begins with the need to build community within the workplace and connect to communities where these businesses are located. And how can this be accomplished? One way is via employee ownership of businesses, creating the kind of presence of place yielding myriad benefits, both within a business and beyond to the larger community.

Employee ownership is a way to strengthen ties between business and community, as it tends to blur the lines of how businesses operate, making them more similar to a social enterprise in many aspects. When employees are empowered as owners, it builds capacity that spills over to the community where employee-owned enterprises are located. Abrams, in working with his own employee-owned enterprise, finds that they attempt to be socially purposeful:

> *by using the financial resources and the web of relationships that derive from our work to help solve community problems and to encourage a better future for the place we live and work. We bring an entrepreneurial approach to these efforts, taking risk and learning from both our public failures and small successes. ... this is the place we know best ... and (we are) doing everything that we can to make a difference in the quality of our community and our economy.*

This commitment to place is an invaluable asset for host communities.

Employee ownership of businesses is gaining recognition and interest across the United States and the world. There are myriad advantages for employees, not only for feeling empowered and part of the company, but also because employee-owned businesses tend to provide better pay and benefits, including retirement. The national organization, Ownership for All, points out the following benefits of an employee-ownership business model:

1 *Preserves jobs and local ownership*. Rather than closing up shop or selling to a competitor, small business owners can sell to their employees. This roots the business and its jobs in the community, and provides a way for employees to share in the wealth created by the business. It should be

noted that most employee-owned firms are conversions, rather than start-ups.

2 *Provides an exit strategy for owners to preserve continuity in the business*. Most owners eventually want to leave the business. Selling to the employees is a way for an owner to exit gracefully, handing over responsibility to managers and ownership to employees at a pace that makes sense for all involved.

3 *Improves company performance*. When employee ownership is combined with participation in decision-making, businesses often see significant increases in performance.

(http://ownershipforall.wordpress.com)

There are two bills proposed by Senator Bernie Sanders that would create a funding source for, and spur the creation of, statewide employee-ownership programs. These bills would be a big step forward in rooting businesses in communities and leading to better quality jobs that spread economic democracy. Communities in the United States will benefit by retaining and growing businesses while at the same time creating more jobs. In this way political democracy would also support economic democracy. The first is the (S.2909), Worker Ownership, Readiness and Knowledge (WORK) Act. It calls for a coordinating Office of Employee Ownership to be created to support local programs for employee ownership and participation. The second bill (S.2914) is called the U.S. Employee Ownership Bank Act, to provide loans and loan guarantees to employee stock option plans and worker-owned cooperatives for purchasing interests in businesses for creation, expansion, and retention of businesses.

Senator Bernie Sanders on employee-owned enterprises: www.youtube.com/watch?v=HDtnp5dlTVE

The value of employee ownership isn't lost on the City of Burlington, or Vermont. Efforts began in the early 1980s by CEDO (see the case study for this chapter in the following Closer Look section). What eventually resulted in a statewide organization began in 1993 with efforts by the Champlain Valley Office of Economic Opportunity, a regional community action organization, and some initial support provided by CEDO. The New Leaf Cooperative Enterprise Program evolved in 2001 into a state-wide nonprofit organization, the Vermont Employee Ownership Center (VEOC). "We promote and foster employee ownership, aiming to retain jobs, deepen employee participation, broaden capital ownership, increase living standards for working families, and stabilize communities, " says Don Jamison, who has spearheaded organizational efforts since 1993. VEOC is a good example of having the employee-owned businesses work together through an organization that fosters their growth and development. Through the efforts of VEOC, a number of firms have decided to become employee owned. VEOC also provides information to a broad range of businesspeople to help them consider if employee ownership is an option to seriously consider.

Figure 2.2 VEOC's logo
Source: www.veoc.org

VEOC is an educational organization focused on promoting awareness about the value of employee-ownership structures by providing information, resources, and technical assistance to owners interested in selling their businesses, to their employees, employee groups interested in purchasing a business, and entrepreneurs seeking to establish new businesses with an employee-ownership model. Currently there are between 30 and 40 employee-owned businesses in Vermont. There are several different forms of employee ownership; here are the two main forms:

1. *Employee stock ownership program (ESOP)*. An ESOP is an employee benefit plan that invests in stock of the sponsoring company. Employees are "beneficial owners" of company stock through a trust. ESOPs are expensive, but there are significant tax advantages for selling shareholders, employees and the company. It's estimated that there are about 10,000 employee-owned businesses; more would develop if more support was in place. Access to capital to transition to ESOPs is one of the biggest obstacles.[8]
2. *Worker cooperative*. A worker cooperative is a business in which the workers are equal owners and have control of major company decisions. Profits are usually distributed in proportion to the number of hours worked in day-to-day operations (source: www.veoc.org). By the way, for more information on cooperatives, look up the Mondragon cooperatives in the Basque region of Spain. It's the world's largest system of employee-owned cooperatives, with over 200 companies employing 80,000 workers. This system is considered the most successful operating cooperative in the world.[9]

Coming back to community

How do all these approaches – programs for small business development, socially responsible businesses, and employee ownership models – relate to the ability to foster a more durable economy? First and foremost is the need for a community to express its desires for long-term outcomes – in Burlington's case it has centered for several decades on a local focus, placing importance on small business development. One of the most engrained ways for community desires to be expressed in Burlington is via its long-term plan for community economic development, beginning with the initial *Jobs & People* report in 1984, and updated periodically. When a plan such as this reflects the vision, values and desires of the citizenry, it provides continuity and commitment and is more resilient against changing election cycles. The basic premise of what a community wants or needs will remain, and while strategies and initiatives may change over time, the desires remain. When an economy has the power of the local guiding it, these desires can be addressed more fully. It's also the perspective that the community is not only participating in the regular or traditional economy, but also the social economy – in ways that generate benefits to the community at large while helping create the place where people want to live, work, play, and visit.

Job creation and destruction

Data on job creation and destruction, or "churning" underlying net employment changes, constitute one very important indicator of the extent to which job development and supporting businesses needs attention. Businesses and organizations start up and shut down, expand or contract, so that jobs are continually created and eliminated. This process of job creation in most economies is dramatic, with jobs created or destroyed in any one year sometimes reaching 10 percent or more of average total employment. At the same time, however, the net change in jobs created overall is relatively small. Unfortunately these data are no longer collected as of 2006 from the U.S. Bureau of Labor Statistics.

Looking at job shift trends from different sectors indicates shifts in the economy. These are how the numbers play out in our region in Vermont (Chittenden County): During a four-year period (from 2002 to 2006) the number of people employed in our region was 94,799; in just a four-year period, 31,167 jobs were created and 30,546 jobs were lost, with the total net new jobs of 621. During this four-year period jobs in the production category decreased by 2,440 and increased in the service sector by 1,828, also reflecting that the jobs disappearing were often higher-paying than the new jobs arriving, which is clearly a concern.

These data point to why a community needs to continually support the start-up of new businesses and the expansion of existing businesses; it is akin to needing to plant new acorns to start new forests for future generations. The above figures are a dramatic illustration of the fact that one-third of the jobs in our region during a four-year period were destroyed or eliminated and replaced with new jobs: that is dramatic, monumental. This is the crux of why a community continually needs to develop its capacity to support this dynamism in the economy through a variety of innovative mechanisms discussed in this book.

(Source: *Jobs & People IV*, pp. 2–15)

Another dimension that's vital for helping support local businesses is finance. Here's a glimpse of what's been done in this area, with some capital ideas through the years at the local level:

- Linked deposit program (City funds deposited in local banks that commit to community reinvestment activities).
- Winter Business Fair to link providers with perspective businesses.
- Established the Burlington Community Banking Council;
- Conducted a credit needs analysis.
- Established a peer lending program.
- Provided counseling to businesses seeking loans, bonding, interim float loans, etc..
- Creation of several loan funds, including a revolving fund and the ONE business loan fund.

Several of the programs and initiatives have been adopted at the state level, including the Vermont Downtown program providing capital to projects in downtowns; development and funding of the Vermont Community Loan Fund business loan program; support of a seed capital fund that the Vermont Center for Emerging Technologies now runs; creation of a low interest loan fund to develop incubator buildings at the Vermont Economic Development Authority; and formation of the Vermont Sustainable Jobs Fund. Some efforts have had influence at the national level – the establishing of a Community Redevelopment Agency group that worked on interstate banking legislation; the conception of the Phase 0 Small Business Innovation Research (SBIR) grant program; the promotion of a 1 percent set-aside from the Community Development Block Grant (CDBG) program for microbusiness programs nationally; and the establishment of the National Association for Community Development Loan Fund (now Opportunity Finance Network) which previously had not been involved with business lending. These programs help to develop capacity at the local level, fostering a more solid grounding for financial capital. Ranging from micro loans to large, complex joint ventures, the capacity to bring capital to bear is critical.

The Closer Look case for this chapter centers on Burlington's work in fostering local ownership. It presents the history of the efforts as well as perspectives on why focusing on employee ownership is valuable.

A closer look: Burlington's Local Ownership Development Project

There are times when Burlington lags behind Chittenden County on key indicators of income and quality of employment opportunities. This gap between the city and the county and related overall trends toward low-skill service employment first came to the attention of the citizens of Burlington and their elected leadership after the release of the 1984 *Jobs & People* report. A Community Advisory Board was formed to look at this issue and the idea of local ownership as a priority was the response. This idea has continued to grow, with locally-focused business as one way to overcome the challenges of maintaining a tax base while serving as a regional economic hub. It's more than just increasing incomes and employment (although these goals are important): "when we build a business, we build a legacy."[10] This especially ties into development of local culture, and the uniqueness of a place when it has locally-focused and locally-owned businesses as part of its community.

The City of Burlington has opted to guide development, embracing an approach that incorporates more than traditional economic development goals. Burlington has to provide the urban amenities for a rural region without gaining the tax revenues of newer growth nodes in suburban locations – this requires a new way of thinking about how to balance economic demands with social and environmental needs. Burlington has recognized that the most direct remedy, in the past and now, is to continue to support existing employers with job training and counseling programs to match resident skills with existing and future job creation. A more long-term innovative initiative is that of the Burlington Local Ownership Project, a special initiative of CEDO designed to encourage the start-up of a range of locally-owned and controlled, for-profit business enterprises and nonprofit and civic organizations – the social enterprises needed to foster well-being and build resiliency.

Starting in 1984, the City of Burlington, working with Chris Mackin and Steve Dawson of the Industrial Cooperative Association, created a long-term economic development framework that focused on local ownership with a preference for employee ownership. The overarching economic development approach focuses on jobs and people and the concept of locally-owned businesses – fusing local business opportunity with employee development. Nearly 30 years later, Burlington is still following this overarching economic development framework, because of a firm belief that this supports and fosters a strong local economy.

The choice of a local-ownership-oriented business development strategy is based on the following assumptions:[11]

- That successful, locally-owned businesses will, over the long-term, provide more stable employment opportunities for Burlington residents since key corporate decisions will tend to be made by residents with a long-term interest in the future health of the Burlington economy.
- That successful, locally-owned businesses will strengthen the local economy as both wages and profits are more likely to be retained and reinvested by local owner/employees.

- That successful, locally-owned businesses, being more familiar with local resources and institutions, are more likely to hire, train, and promote local residents, therefore promoting a higher percentage of quality job opportunities in the community.

Further, CEDO states that employee-owned and -controlled businesses should be particularly encouraged because of:

- Their demonstrated performance potential. Studies have found employee-owned businesses outperform conventionally owned business structures on measures of productivity and profitability.
- The breadth of local ownership which they can provide – in placing long-term strategic decisions, that could affect the local economies, in the hands of a broader number of local actors than one or two local entrepreneurs.
- The quality of the employment environment they can create by involving local residents in decisions which affect companies that they own.
- The fundamental equity and fairness of employee ownership as a business structure – which helps distribute the gains of economic success to the people most responsible for that success – the blue-, white-, and green-collar employees working under the same roof together.

Within this local ownership business development strategy, a number of enterprise structures are supported, including: (1) Individually owned, entrepreneurial start-ups – possibly taking advantage of City-sponsored incubator business space; (2) Family-owned entrepreneurial start-ups; and (3) Employee-owned and -controlled businesses.

Implementation

Actual strategies for encouraging local ownership of business can be summarized under three basic categories:

1. Import substitution start-ups where new, locally-owned business is started to produce a product or a service which major local employers must presently import from out of state. Major employers stand to benefit from these enterprises through the provision of a ready, convenient source of supply which would reduce their need to carry excess inventory. New, locally-owned enterprises would receive a temporary, sheltered, local market which can assist a company in its early start-up stages.
2. Conventional, entrepreneurial start-ups where local entrepreneurs proceed to organize a locally-owned business on the strength of a new product or service idea designed for a variety of markets – local, national and international. Likely sources of new business ideas could include university research and development centers, oriented towards encouraging local ownership.
3. Conversions of retiring owner businesses where existing healthy, local businesses find no likely or desirable conventional outside buyer and where local, internal management or management/employee groups move to purchase the firm themselves.

Various financing and technical assistance structures exist to help entrepreneurs interested in developing locally-owned businesses, including the Burlington Revolving Loan Program, several socially conscious investment funds, the Cooperative Fund of New England, the ICA Group for community lending, and the Vermont Employee Ownership Center.

Gardener's Supply is one example of a company which has benefited from the local ownership approach and which in turn has served the Burlington community. CEDO began working with them over 25 years ago when they were still a small company. Gardener's Supply has since grown tenfold, and as of December 2009 sold 100 percent of the company to its workers. Burlington's and Vermont's creative and independent nature helped fuel a decision, made 25 years ago, for Gardener's Supply to invest in their employees. In 1987, after their third year in business, an Employee Stock Ownership Program (ESOP) was adopted that allows all employees to earn stock and share in company profits. Employees are encouraged to learn about the entire dimension of the business, empowering them to actively participate in guiding current practices and future outcomes. They found that by staff serving as owners, their creativity and commitment needed to be responsive to customers was enhanced, as well as sustaining a vibrant focus on gardening and cultivating a compassionate corporate culture in which all are rewarded, including the community in which they operate.

It isn't an easy process to convert to employee ownership, and it requires a good deal of adjustment and flexibility. But it can yield benefits to both the company and the surrounding community. Will Raap, the founder of Gardener's Supply, can attest to that. "It's a commitment to something bigger than yourself," he explains,

> *it's also about looking at different capitals a community has; Burlington does a good job of thinking this through and not yielding to quick fixes or quick buck solutions from outside. Instead, it's driven by asking "what will the economic development drivers be such as a local business focus?"*

About that commitment to something bigger than one's self, it should be noted that in 1986, Raap spearheaded an effort to restore a floodplain area in Burlington along the Winooski River, converting it into a 350-acre farm incubator breadbasket. The Intervale Center is an educational and production center for the city

Figure 2.3 Gardener's Supply retail store located in Burlington's Intervale
Photo: Gardener's Supply
Source: Will Raap, founder, Gardener's Supply.

and region, producing over 1 million pounds of food per year and serving as an agricultural showcase project (see Chapter 7 for the case study included in the Closer Look section).

"The positive ethic around business responsibility in Burlington led us here," says Raap.

Resources and ideas for making it happen in your community

See the latest *Jobs & People* (2010) at the City of Burlington website, www.burlingtonvt.gov/cedo/

Business Alliance for Local Living Economies: www.livingeconomies.org/

Certified B Corps: www.bcorporation.net/about

The Employee Stock Ownership Program Association: www.esopassociation.org/

Gardener's Supply: a company of gardeners, rooted in Vermont and 100 percent employee owned

Gardener's Supply is in business to spread the joys and rewards of gardening, because gardening nourishes the body, elevates the spirit, builds community and makes the world a better place.

(Mission statement, 1984)

Gardener's Supply was founded in 1983 by a handful of enthusiastic Vermont gardeners. Today, we serve millions of gardeners nationwide, offering everything from seed starting supplies and garden furniture to flower supports and garden carts. Though our company has grown, we remain passionately committed to providing garden-tested, earth-friendly products that will help our customers have more fun and success in their gardens.

Gardener's Supply is proud to be employee owned. We are gardeners ourselves and have earned our customers' trust by providing high-quality products, expert information and friendly, personalized service. At our headquarters in Burlington, Vermont, you can *check out our store* and stroll through our *3-acre display gardens.* We also offer local seminars and special events throughout the year.

Improving the world through gardening

Gardener's Supply actively promotes gardening as an important way for people to make the world a better place. *We donate 8 percent* of our pre-tax profits to support programs and organizations that are using gardening to improve the quality of people's lives and the health of our environment.

A cornerstone of our donations program is the *Garden Crusader Awards*, which recognize people across the country who are using gardening to improve the quality of life in their communities. Each year we select 21 winners and award more than $15,000 in cash and gardening products.

Gardener's Supply also provides gifts to a number of nonprofit organizations working on gardening, sustainable agriculture, the environment and hunger-related causes. Recipients include the American Community Gardening Association, the American Horticulture Therapy Association, and Plant a Row for the Hungry.

In Vermont, our company's local donations program helps support more than 50 community organizations. We also founded, and continue to be a lead sponsor for the *Intervale Center*, which oversees 350 acres of farmland and a wide range of urban farming initiatives in Burlington, Vermont.

We are mindful about our company's environmental responsibility, and are pursuing a number of environmental initiatives to reduce our carbon footprint. To learn more, read *Our Commitment to the Environment.*

(www.gardeners.com/Who-We-Are/5445,default,pg.html)

National Center for Employee Ownership: www.nceo.org/
Vermont Businesses for Social Responsibility: http://vbsr.org/
Vermont Employee Ownership Center: www.veoc.org

Individual business development strategy: If you're in business and have plans to grow, call your community's planning and zoning department to have a conversation about your project early on in the process.

Founders' board: If you are starting a business or running a business, establish a founders' board of trusted advisors to help you prepare, evaluate, and discuss your business operations. Include people with diverse perspectives who will ask you tough questions. Meet at least three times a year, more often when you are first starting out.

Tax incentives: Research federal and state tax credit incentives for a variety of needs, including access for people with disabilities, research and development, historic restoration, hiring, energy efficiency, brownfield restoration, carpooling and transportation for employees. These incentives coupled with a good business plan can help a business grow.

Local retailers operate like stores in a mall: Work with local retailers to establish and post uniform shopping hours, like in a mall. Shoppers will be happy you did.

Research local business needs: Develop or buy list of businesses in town. Survey manufacturers and value added businesses to identify potential retiring owners. Find out what needs are not being met.

Business security planning: If a business has chosen a location to start or expand, advise management to meet with the police and fire departments to get their advice on safety and security issues.

Climate adaption: If a business is locating in an area that is near water it is critical to assess risk and consider the impact of flooding on business operations.

3 Welcome One, Welcome All: Economic Inclusiveness

How Are All – Women, Minorities, New Arrivals, Disadvantaged, and Others – Integrated into the Economic and Social Fabric of the Community?

Inclusiveness is smart business and development strategy. Microenterprise development, women's business development programs, activities to integrate members of immigrant communities, rehabilitation assistance for new skill development and new business development – these and other initiatives build a local economy accessible to all.

The idea of a community being a good place for all – not just for a selected few – is key. It's about doing a good thing, or to put it another way, it "is not about the conviction that something will turn out well, but the *certainty that something makes sense, regardless of how it turns out.*"[1] These actions "may be just a drop in the bucket, but rivers are nothing more than accumulations of drops ... each drop has an impact."[2] It's also the idea of fostering an inclusive economy sector, one in which the right thing to do is a good thing to do, with benefits for all. This impacts a community's ability to foster a durable economy, which is impossible if segments of population are left out.

Inclusiveness of all in a local economy is not only a good thing, it makes very good sense from the economic development vantage point. There are times when larger, existing companies contribute to job destruction in economies, losing employees as cyclical changes within their industry sector occur. On the other hand, small businesses, including microenterprises and small companies with less than 500 employees, tend to generate new job growth in the U.S. economy each year. Burlington and CEDO decided early on that the focus would be on small businesses, rather than chasing the larger companies that may or may not be sustainable over the long term. Since the beginning in 1984, the strategic economic development plan, *Jobs & People*, has centered on ways to encourage small business development, including for women and minorities – those typically not fully represented in most economies. This chapter focuses on two areas: the first for skill and business development for women; and the second for integrating newcomers, immigrants and refugees into the community. The Closer Look case study in this chapter focuses on the Vermont Refugee Microenterprise Program.

Occupations and gender

In the latest *Jobs & People* report, there are quite a few differences by gender for certain occupations in Burlington. Using 2008 data, it shows that women dominate in the service sector by more than 10 percentage points, as well as in sales and office, and managerial, professional, and related occupations. On the other hand, men hold the majority of farming, construction, and production positions. These results are interesting because while the majority of the shrinking fields are male dominated, women in those fields have fared worse than the men have. In the women-dominated fields, there has been a mix. Women have taken more than two out of every three new service jobs since 2000. Managerial, professional, and related jobs have also held steady for women, while male employment has fallen by nearly 10 percent. At the same time, women have fared worse than men in sales and office positions, with female employment falling by 23 percent.[3]

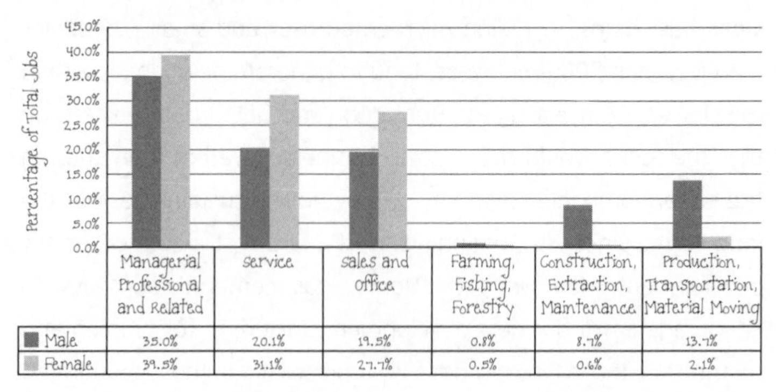

Figure 3.1 Employment share by occupation and gender, Burlington, 2008
Source: U.S. Census Bureau, American Community Survey 2006–2008, 3-year estimates.

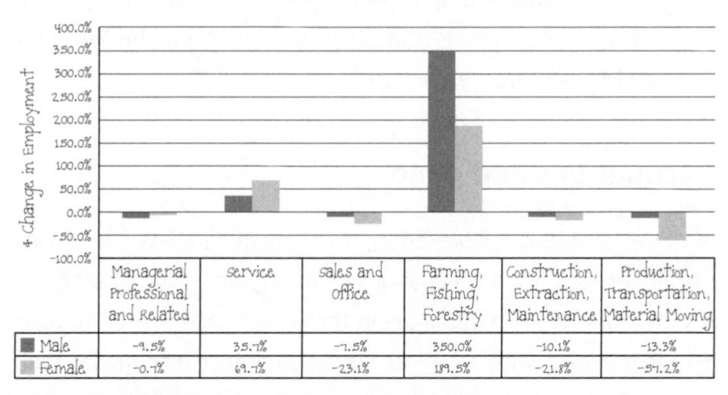

Figure 3.2 Percent change in employment by occupation and gender, Burlington, 2000–2008
Sources: U.S. Census Bureau, American Community Survey 2006–2008, 3-year estimates. U.S. Census (SF3) Series PO50, 2000 data.

Women at work

A focus on inclusiveness had its inspirations in findings from the City of Burlington's 1984 *Jobs & People* report. It was clear that groups existed for whom the region's prosperity was merely someone else's good news, particularly low-income single female heads of households. From this initial prompting, several actions were taken, leading to job development and business development programs for women. These actions have led to a variety of outcomes, including over 140 women-owned businesses in the area, the Women's Small Business Program of Mercy Connections and a spin-off, state-wide program, Vermont Works for Women.

How? "Go do it," has served as CEDO's modus operandi since its inception. Martha Whitney joined CEDO in 1986 as director of the new Women's Economic Opportunity Program, and proceeded with this charge of getting things done. "We started by focusing on the need to support low-income women for developing skilled trades, helping them transition off welfare, and being able to more fully participate in the economy," she explains, "it was an era of new progressiveness, and we believed that we could do it." Models around the country and beyond were assessed for ideas and best practices, and local needs analyzed. Several key considerations emerged:

1. *Collaboration is vital. Find how to support each other. Collaborating with an existing Women's Council started the ball rolling.*
2. *Commitment is needed on all levels, not only at the community level but also organizational level with backing from elected officials to staff.*
3. *Leverage outside resources with activities such as grant writing to bring in funds to run the program. The first funding came from the Vermont Department of Labor's Job Training Partnership Act (JTPA), formerly the Comprehensive Employment and Training Act (CETA) and Burlington's Community Development Block Grant (CDBG).*
4. *Hire competent staff – this is essential: creative staff leaders make it happen.*

It was the high quality nature of collaboration that enabled this original initiative to spawn many other programs. Quality of these collaborations and relationships is vital, and serves as the foundation for building social capital capacity in any community. Commitment was critical too, from all levels of the organizations involved. Leigh Steele, who was involved in the early efforts to form Vermont Works for Women, notes that "collaboration is key, so much happens because we work together with the State, nonprofits, County and others; it is also crucial to get the community involved."

In assessing the economic outlook for single mothers in the 1980s, it became apparent there was a need to support women in the skilled trades. The trades typically pay more than other jobs and demand at the time was strong. The response started with the Step Up Program focused on providing trades training in three different areas (electrical, carpentry, and plumbing) over 13-week periods. Lots of education was included, and not just for trades skill development. The program encompassed fitness training (to build strength and health), resources for managing other aspects of life (such as financial management), and assertiveness training (for helping overcome barriers). At first, the male-dominated contractors' associations were resistant, but Mayor Sanders backed the Step Up Program. A new City initiative requiring contractors on City projects to have 10 percent minority (gender or racial) workers remedied some of the immediate resistance.

Vermont Works for Women

The Step Up Program continues and has proven to be a very empowering initiative for its participants, with lots of follow-up support. It influenced creation of the Northern New England Tradeswomen organization, and grew into Vermont Works for Women, a 501(c)3 nonprofit organization with expanded opportunities including women police officer training and job rehabilitation training for women in prison. Their mission is to assist women and girls to help them develop skills and capacities critical to long-term economic independence. This is not only good for women; it's also good for their families and for the economy.

Figure 3.3 CEDO's Guide to Doing Business in Burlington

The award winning *Guide to Doing Business in Burlington* is one of many CEDO publications for the business community. The 196-page Resource Guide for Chittenden County Employers and Employees is available online, as are the Business Refugee Resource Guide and Business Location Information. CEDO also maintains and distributes a free commercial space database averaging close to 100 listings throughout the city and provides one-on-one assistance to businesses and entrepreneurs. For a pdf download, see the link at: www.cedoburlington.org/business/doing_business_in_burlington/TheGuide.pdf

Because nearly one-third of single mothers in Vermont live in poverty (many despite working full-time), it is crucial to help women and girls realize their potential for careers paying a living wage. In addition to women's skill development programs, Vermont Works for Women offers support for girls, with Rosie's Girls Summer Program – Building Strong Girls for junior high students; a high school program, Women Can Do Conference; TechGrlz provides an alternative education program for high school girls in partnership with the Center for Technology in Essex and the Vermont Department of Labor; and gender equity consulting for schools.

Funding has grown from City support, to state and federal with the U.S. Department of Education providing grants. Several participants in these programs have founded their own businesses, providing services in areas typically dominated by males. Girlington Garage is one example, serving fair trade coffee, providing wireless internet access and free shuttles in a comfortable environment for auto repair.

Here's a summary of the programs currently offered by Vermont Works for Women, aimed at helping women enter nontraditional careers with a livable wage.

- Step Up to Law Enforcement for Women – *an innovative partnership with nine law enforcement agencies that includes a full-time, nine-week training program for careers in local law enforcement and with the state's police and corrections academies.*
- Step Up to Carpentry – *This is a six-week program with carpentry, math, and employability skills for entry into commercial and residential carpentry careers.*
- Step Up to Painting for Women – *Developed in collaboration with the Green Mountain Chapter of the Painting and Decorating Contractors of America, it provides a six-week training program for entering painting careers.*
- Building Homes, Building Lives – *This program for incarcerated women focuses on the building of modular homes that are installed as affordable housing units in Vermont communities. It prepares women for employment in the construction trades through training, job readiness skill development, and a portfolio of work the participants can show prospective employers when transitioning back into communities.*
- Northern Lights House – *Providing a supportive residential living environment for women under correctional supervision, the program emphasizes*

individual and group responsibility to support living healthy, productive lives.

- Vermont Women's Mentoring Program – *In partnership with Mercy Connections and the Vermont Department of Corrections, this program assists women as they transition back into their communities. Matching female volunteers from the community with women being released from prison for nonviolent offenses offers a support mechanism for transition.*

Leigh Steele, who has been involved with the City (CEDO) and Vermont Works for Women since its inception, points out that the modular housing program has garnered national attention. "This program is helping women succeed, and lowers recidivism rates dramatically," Leigh said. The recidivism rate of 19 percent for the Building Homes, Building Lives program graduates over the two-year time period 2006–2008 is less than half the recidivism rate for the general female prison population (51 percent).[4] In other words, the program lowers a participant's chances of ending up back in prison.

Women's Small Business Program

Translating skill development into business ownership is a tremendously beneficial way to help integrate women into a local economy. Small business development for women was next in the quest to overcome disparities. In 1989, the City turned to a local institution, Trinity College, offering the Women's Small Business Program (WSBP) via a cooperative effort with CEDO. In the early twentieth century, the Sisters of Mercy Vermont established Trinity College; when it closed in 2001, they wanted to continue their legacy of helping the communities they served and established Mercy Connections as well as the Institute for Spiritual Development. Mercy Connections, a 501(c)3 organization, is the educational arm of the Trinity College legacy, that in the spirit of compassion, justice, and hospitality supports those in need. They have dual priorities of acting in solidarity with those who are economically disadvantaged, especially women and children; and working together, using combined voices in the wider community, to change the social and legal policies that hinder individuals from reaching their full human potential.[5] Mercy Connections continues the work of the WSBP; over 2,000 women have participated in this program since 1989. The purpose is to provide skill building, mentoring, and peer support to position Vermont women for success in business.

WSBP offers the following programs:

1. *Getting Serious, a workshop to explore the possibilities and realities of business ownership.*
2. *Start Up, a 15-week intensive program for women who are starting, stabilizing or expanding a small business, as well as those who would like to test business idea feasibility.*
3. *Working Solutions, ongoing training seminars and individual counseling for business success.*

It is important to note that slightly over half of participants are qualified as low income. This reflects not only need in the community, but also the mission of Mercy Connections to support women in poverty and transition by empowering them to make positive and long-term changes in their lives for economic well-being. While women account for almost 50 percent of Vermont's workforce, their average annual earnings total 65 percent of men's yearly earnings. Challenges such as lack of education, time flexibility, and the gender wage gap contribute to this.

Over 140 new women-owned businesses have started in the area with the encouragement of the WSBP. One of the keys to starting a new business, according to Betsy Ferries, a former director of the program, is to help build relationships and connections in the community – these local connections provide assistance with critical areas such as business plan review and identifying financial resources. "We offer intensive marketing, personal and professional development coaching to help overcome challenges to starting and succeeding in business," Betsy explains. Enabling confidence building and support provides a winning combination.

In the 2009 WSBP Directory of Graduate-Owned Businesses, 12 categories of businesses are listed, from agriculture, artists, computer services, health and wellness, professional services, and publications to travel and vacation.

At the risk of being reiterative, the success of these programs has hinged on the willingness to build relationships, collaborations and partnerships. One organization alone can't accomplish all this but working together is enabling. As stated early on, social capital is built in communities via relationships. These initiatives and programs exemplify the role of social capital in fostering inclusiveness, yielding results that positively impact residents and help strengthen the economy.

Women's Small Business Program

See the following video for perspectives from participants in the program: www.mercy connections. org/php/Videos. php

---

The Minority Assistance Program (MAP), with a five-year grant to CEDO, provided technical help to disadvantaged businesses, including a resource library, seminars, referrals, and the Minority Business Forum. Supported by a U.S. Small Business Administration Grant, MAP has raised awareness of the minority business population and the challenges these businesses face.

Resettling refugees

Burlington's role as Vermont's first political refugee resettlement community began in the 1980s. Later, then mayor Peter Clavelle was asked whether or not Burlington, as a newly designated refugee city, would take a group of Bantu people from Somalia. More established refugee-designated cities had already turned them down with the justification that they didn't possess enough skills or education needed in the resettlement process. "Of course, why wouldn't we" (a statement, not a question) was Peter's reply, and Burlington launched headfirst into its new role. At first, there wasn't a refugee resettlement office, but with a three-year AmeriCorp*Vista position, the City began to garner skills, resources and acumen to respond to resettlement challenges. As a result of working with the refugees, funding from the state was obtained to organize this program via VISTA. The City developed a network, and then worked to obtain funding for a state-wide position to coordinate all efforts. This position is now served by a state employee. Neighboring towns in Vermont are now also refugee communities (Winooski, South Burlington, and Barre).

The initial contact for most refugees in the state is the Vermont Refugee Resettlement Program (VRRP) located in Colchester, a town adjacent to Burlington. This is a program and local field office of a national nonprofit, the U.S. Committee for Refugees and Immigrants. Here, refugees begin the resettlement process in Vermont. VRRP provides a variety of services including case management, English for Speakers of Other Languages (ESOL) instruction, professional interpretation/translation, job placement, social adjustment/mental health counseling, family reunification, and other federal assistance available to all refugees settling in the United States. The objective is to help refugees achieve economic and social self-sufficiency within eight months, before the federal assistance ends. The VRRP resettled 6,800 refugees in Vermont, with Burlington resettling close to half of these from Bosnia, Vietnam, Congo and other nations from 1980 to 2012. The program has grown, and Vermont has settled more than 330 refugees annually on average over the past five years.

With a significant number of refugees currently living in the Burlington area, it is critical that a broad network of service providers be maintained. CEDO works to achieve economic justice and sustainability in Burlington by mobilizing resources and working with businesses, schools, nonprofits and citizens to increase economic opportunities for all, and particularly to extend services to low-income and disenfranchised populations. The Refugee Business Resource Guide was created in response to Burlington's growing refugee population and the need for local employers to address challenges to hire, retain, and promote refugee workers.

Figure 3.4 Martin Luther King celebration at City Hall in Burlington
Photo: CEDO.

CEDO recognizes this unique and valuable segment of Burlington's workforce and seeks to collaborate with the many companies and small and emerging businesses that support refugees. This guide is a step in linking the business community with the vital service providers to promote the economic self-sufficiency of refugee workers. Many services such as children and family services, health care, ESOL, and housing are integral to the immersion into American life. The Business Refugee Resource Guide[6] was completed as a joint project of CEDO and AmeriCorps*VISTA, in cooperation with five Burlington businesses that inspired this project: Burlington Furniture Company, Conant Metal and Light, Inc., Fresh Connections, Lake Champlain Chocolates, and Rhino Foods.

Building capacity

The Champlain Valley Office of Economic Opportunity's Chittenden Community Action Agency, in conjunction with community partners, initiated the Microbusiness Development Program (MBDP) to provide region-wide business technical assistance to low- and moderate-income residents and identify opportunities for refugees to participate in the local economy. The Micro Business Alliance, comprised of members from CEDO, MBDP, Small Business Administration (SBA), banks and credit unions, University of Vermont faculty, Community College of Vermont, business consultants, Small Business Development Center (SBDC), Women's Small Business Program and others, meets on a quarterly basis to discuss the condition of the local and regional economy and to find and implement opportunities. One of the most successful to date is the ONE World Market, an outdoor global-style event reflective of the traditional marketplaces in other countries. ONE refers to the Old North End neighborhood of the City. The idea of the market is to provide a place to test the business ideas of refugees and immigrants offering products or services. CEDO and MBDP provide microenterprise training to prepare the vendors for the market experience, including retail display advice, professional merchandising assistance, and customer interaction training. Most of the vendors do not operate formal businesses or have a business plan; rather, the market provides a safe venue to test ideas and products in an environment where it is easy to mix and meet others in the community. A few vendors do operate established businesses and this mix of experience provides an "open air incubator" environment for sharing expertise among the participants. Translation services provided by the Vermont Refugee Resettlement Program were very helpful during these activities.

Due to tax implications, these events must have 24 vendors or less. Seven ONE World Market events have been held in the period from 2007 to 2012. One of the markets was held in conjunction with the annual South End Art Hop, one of the biggest special events in the city (see Chapter 4), as well as various locations around the city. The Microbusiness Development Program also held an indoor holiday market during the winter, where much was learned, including the need for vendors to acquire insurance.

Several refugees have started businesses from their experience with the ONE World Market. These include businesses providing jewelry, food, metal working, dishes, clothing and cloth, and soap. Microbusinesses are important for the refugees and immigrants who may be in need of ways to earn supplemental income and connect to others in the community. The downtown local City Market carries some of these items to promote locally produced products (see Chapter 6).

Building relationships and connections is vital for the refugee and immigrant population and women, as is business and skill development. To help, the City invites service providers to the events, such as banks and credit unions, to meet with refugees. The markets and training events provide a safe place for meetings and conversations with others that otherwise may never happen. This safe and comfortable environment helps overcome the reticence of refugees to go into banks, for example.

One of the keys to building capacity is to engage lots of community organizations, groups such as neighborhood based groups and others – basically, pull in as many as willing to help build capacity. Organizations have formed, such as the Association of Africans Living in Vermont, that help bring in individuals who otherwise may be isolated. For example, this group has proposed the idea of drawing in African women's indigenous arts and skill sets such as batik. Initial ideas for potential markets for the products include making eco-friendly shopping bags using beautiful batik designs on natural materials.

Lessons learned about the ONE World Market include the following: different refugee groups selling in the markets have an opportunity to reach out and not remain isolated; the markets provide a venue for transactions, whether personal or business, by creating a chance to interact with different cultures that may otherwise not normally occur; the opportunity to talk with bankers and other service providers; and the ability to meet people as potential customers and distributors in a safe setting is critical. Long lasting benefits are created when these opportunities are provided. After the market, August First Bakery opened, hiring several employees through a refugee community social services organization.

The willingness to do what it takes to build capacity and respond to challenges is crucial. Again, the ability to work across organizations, cultures and perspectives provides the opportunity to bring refugees and new arrivals, both men and women, into the economic and social fold of a community – those who may otherwise remain separate.

The importance and value of including all and creating processes and programs, as well as a community culture to accomplish this, has been shown. It's no longer an option to just say that a community or economy is open; it needs to be demonstrated through action and results. It has to be accessible for all those who live in a community. It takes a collaborative and concerted effort across organizations to be effective. Fostering a supportive community environment requires an attitude of acceptance backed by resources.

The case selection for this chapter focuses on the Vermont Refugee Microenterprise Program because it illustrates both collaborative effort and willingness to move beyond a typical small business service menu. If this is done an economy can grow, as witnessed by 140 new women-owned businesses established with the support of the Women's Small Business Program since its inception.

A closer look: the Vermont Refugee Microenterprise Program[7]

Now completed, this program provided insight into the process and results of constructing and providing microenterprise development assistance to refugee entrepreneurs. This case looks at a three-year U.S. Office of Refugee Resettlement grant period, from 2002 to 2005, documenting results from the program and initiatives. The Microbusiness Development Program was created by the Champlain Valley Office of Economic Opportunity's Chittenden Community Action Agency in collaboration with the City. Why? More attention was needed to help refugees transition into the economy, beyond the assistance provided by the federal and state programs (assistance from federal sources ends eight months after arrival in the U.S.). It included three different business counselors who provide services to refugees; by offering these programs, the Microbusiness Development Program built its capacity to work with these populations. It also is a result of several individual organizations that all contributed to these efforts.

The refugee clients included the following ethnicities:

- 1 from Afghanistan.
- 12 from Central Africa, Sudan, Burundi, and Rwanda.
- 17 from Vietnam.
- 26 from Eastern Europe, including Bosnia, Azerbaijan, and Moldova.
- 32 from Somalia (all Somali-Bantu refugees).
- 39 from the Democratic Republic of the Congo and the Republic of Congo.

The ethnicity of the clients mirrored the ethnicity of the refugees resettled in Vermont by the Vermont Refugee Resettlement Program. Initially, there were a number of clients from Bosnia and Vietnam – those refugees who had been resettled for longer periods of time. A number of clients from these countries of origin found out about the program via word of mouth and were no longer eligible for the state's Refugee Program, as they had already become citizens. However, it was possible to provide them with services through the regular MicroBusiness Development Program.

Members of these communities started several businesses. Bosnian and Vietnamese Bilingual/Bicultural Counselors (B/BC) assisted these clients with the help of the business counselor. The second flow of clients came from the two Congos and the other African countries (excluding Somalia). Most of these clients entered the program as a result of the outreach efforts of the Congolese B/BC. Many attended training sessions. A number were enrolled in the Financial Literacy and Individual Savings programs. Members of this community started several businesses and were assisted by the business counselor. The most recent arrivals were Somali-Bantu from Somalia. Although 32 were enrolled in the program and a number attended training sessions, the only businesses started in this community were home-based childcare businesses. Somali interpreters and the business counselor assisted these clients.

The following table lists the program goals and outcomes, with any special considerations listed in the far right column.

Table 3.1 Refugee Resettlement Program goals and outcomes

Program goal	*Outcome*	*Comments*
Enroll 70 refugee clients	Enrolled 127 refugee clients	
Deliver training to 40 refugee clients	Delivered training to 91 refugee clients	Many clients attended multiple training sessions. The materials created for Financial Literacy will be used by the Vermont Refugee Resettlement Program.

Provide intensive pre-loan and credit counseling to 24 refugee clients	Provided intensive pre-loan and credit counseling to 26 clients	Several clients were able to establish and/or repair credit. Several participated in financial literacy and savings programs and were able to use the combined savings/matched grant to start businesses. See note on Individual Development Accounts (IDAs).
Assist in the start-up of 15 new refugee business start-ups	Assisted in the start-up of 44 business starts	See note on Types of Business.
Create 37 new jobs	Created 39 full time and 41 part-time jobs for a total of 80 jobs	Some part-time jobs are seasonal.
Stabilize 5 refugee businesses	Stabilized 6 businesses	
Provide 10 First Step loans to refugees	Provided 8 First Step loans	Applied for 10 First Step loans but two were not approved due to major credit issues. These two clients received credit counseling. Five refugee clients who received First Step loans were able to leverage these loans to obtain additional financing for their business starts.

Table 3.1 continued

Assist refugee clients to obtain business loans	Assisted refugee clients to obtain $284,983 in outside business loans	Larger loans were from traditional banks; smaller loans were from nonprofit organizations such as Vermont Job Start and the Opportunities Credit Union.
Assist refugee clients to obtain grants	Assisted refugee clients to receive $10,700 in grants	See note on Individual Development Accounts.
Assist refugee clients who wish to pursue employment	Assisted 37 clients to pursue employment	Assistance included help with resumes, online searches for job opportunities, and assistance at job interviews.

It is notable that most goal targets were exceeded, indicating both effectiveness of program design and delivery as well as need in the refugee community. The participants' responses to the program were positive, with several notable outcomes including enhancing the ability of refugee-based businesses to obtain outside business loans.

Training program

Training materials were translated into French, Bosnian, Russian and Vietnamese, and were delivered in English with an interpreter. The materials were not translated into Somali because the majority of the Somali-Bantu clients were unable to read; in those cases, interpreters were used to deliver the information via workshops. These materials covered a variety of basics and specialized topics to support business development, including: money and credit basics; business in Vermont; basic bookkeeping; business taxes; effective marketing; food production; childcare; home ownership; computer literacy; and financial literacy for literate and nonliterate clients.

Computer literacy program

The MicroBusiness Development Program leveraged the Office of Refugee Resettlement grant with two additional grants – from Vermont TECHCorps and the Verizon Foundation. These grants provided a computer laboratory and instructor for 2005, coinciding with the third year of the refugee grant. Many of the refugees were incorporated into this training and mentoring program. The B/BCs assisted in the training and the software was set up to support the refugee clients. Some highlights of this program included:

- A computer laboratory was set up with five computer systems loaded with standard MS Office, voice recognition, and Typing Tutor software. All systems had access to high-speed internet, a laser printer, and a scanner.
- A qualified former Bosnian refugee ran the computer lab and provided classroom training and mentoring.
- 12 volunteers (including 6 refugees) assisted with the training and mentoring.
- 57 refugees received training in computer literacy.
- 37 refugees completed computer basics.
- 24 refugees completed Word training (5 at an advanced level).
- 5 refugees completed Excel training.
- 15 refugees completed Windows training (5 at an advanced level).
- 11 refugees completed email training.
- 10 refugees completed internet training.
- 1 refugee completed PowerPoint training.
- 18 refugees used the computer lab on a regular basis to practice typing skills.
- 15 refugees linked their computer literacy training directly to self-employment goals. Of these, 8 started businesses.

- Youth training sessions were held in the summer for Somali-Bantu and Congolese youth. A number of teenagers attended, including several Somali-Bantu teenage girls. Advantage was taken of their attendance at computer literacy training to also provide them with some basic financial literacy training relating to employment and savings.
- Two "refugee computer parties" were held, with over 60 refugees attending; a lottery was held at each event to give away 12 computer systems.

Pre-loan and credit counseling (IDA – Individual Development Accounts)

Individual pre-loan and credit counseling was provided as part of the Refugee Microenterprise Program. This program is housed within the MicroBusiness Development Program of Chittenden Community Action, providing technical assistance and training to low- to moderate-income residents who either already operated a business or wanted to start a small business. One-on-one counseling was available, including business development services such as creating a business or marketing plan and securing funding. Expansion of services focused on personal financial classes and asset development strategies.

Chittenden Community Action

This group serves as a facilitator for the area, bringing together small business groups such as the Small Business Development Center and others quarterly to focus on needs and actions. They were also able to leverage an existing program to provide financial literacy training to a number of refugees, with the help of the B/BCs who provided interpreting. The participants in this program were also able to contribute to Individual Development Accounts.

Qualified applicants saved towards business starts, home ownership, or education. Upon completion of the program, including monthly deposits into a savings account for the required period, the amount saved is matched by a grant. It's a way for aspiring entrepreneurs to leverage meager resources. The results for the refugees enrolled in this program were:

- 14 refugees completed the training and participated in the savings program.
- 2 refugees withdrew their savings before the completion of the program due to financial emergencies.
- 12 refugees saved a total of $11,041.98 by December 31, 2005.
- The total funds available (with matching grants) were $24,248 as of December 31, 2005.
- 5 refugees used their IDA money to start businesses.
- 1 refugee used the IDA money towards buying a house.
- 2 refugees used the IDA money for education.

Types of business

A variety of businesses were started during the program. It is clear that the majority of these businesses would not have been started without the assistance of the Vermont Refugee Microenterprise Program's team of bilingual/bicultural counselors and business counselors, who in many cases were also able to provide emotional support because of their sensitivity to cultural issues.

Retail

Several retail businesses were started including three grocery stores, one convenience store/gas station, a clothing business, and one fabric business. All three of the grocery stores carry ethnic foods – one carries European/Middle Eastern foods and the other two carry foods that are popular with Africans. The Vermont Refugee

Microenterprise Program assisted all of these businesses with business plans and loan applications, as well as the legal and business requirements for retail businesses in Vermont. At least one of the business owners leveraged the IDA savings program and a First Step loan to get started. Other services included help in finding locations, help with leases, assistance in finding and negotiating with suppliers, and assistance with marketing.

The grocery stores are in Burlington; CEDO helped find locations and get these businesses set up. Most of the prospective owners had a very limited command of English at the time they started. The no-cost extension period for the grant was very useful in helping to stabilize these businesses and take them through the end-of-year formalities.

Food

A number of food businesses were started. A Congolese refugee started the most successful. Samosa Man makes and sells samosas with a variety of meat, vegetable, and fruit fillings. It's interesting to note that a more established entrepreneur in the area invested in this start-up business: Ben Cohen of Ben and Jerry's. The products are sold at farmers' markets across the state and also through specialty food stores, and even in the school lunch program. The business owner received considerable assistance from the Vermont Refugee Microenterprise Program, which introduced him to Vermont Food Ventures and found him an accountant. The program also assisted with marketing, finding a location, business plans and loan applications.

Other food businesses include: Russian pastries, Mid-Eastern delicacies (Pakistani foods), egg and spring rolls, and specialty breads. Some of these foods are sold at farmers' markets and some through grocery stores, including local cooperatives.

Cleaning

Bosnian refugees, focusing on commercial cleaning, started several cleaning businesses. One business gained a contract to clean the rest areas along the interstate highway. One business focused on medical and professional buildings. Refugees from Azerbaijan and the Congo started other cleaning businesses.

Only one of these businesses required a loan to get started and the business owner was able to leverage a First Step loan into another, larger loan from the Opportunities Credit Union. All businesses benefited from the services of B/BCs and business counselors because the business owners' command of English was limited. Marketing materials, including brochures, flyers, and business cards, were created for some of these cleaning businesses.

Childcare

Vietnamese and Somali-Bantu women started several childcare businesses. These are all home-based childcare businesses that provide incremental income for the women. Most of the women attended financial literacy workshops that were developed for non-literate clients. The no-cost extension grant period was used to stabilize some of these businesses and to help the women complete their year-end records in preparation for their income tax returns. It was also used to review the program with the Vermont Refugee Resettlement Program and to educate case managers and staff on the program, so that they could take over providing support services to some extent.

Restaurant

A refugee family started a Vietnamese restaurant. Due to a lack of English on the part of the prospective business owners, the Vietnamese B/BC's assistance, which focused on the legal and business regulations and requirements, was essential to this endeavor.

Transportation

A refugee from the Congo started a taxi business. The Vermont Refugee Microenterprise Program assisted with marketing materials and a website, as well as with business counseling. This business has been successful and expanded to a fleet of five taxis.

A refugee from Bosnia started a trucking business. With assistance from the Vermont Refugee Microenterprise Program, the individual was able to create a business plan and obtain a loan from the Opportunities Credit Union to buy the first truck. He has been so successful with this business that he has been able to expand by buying a second truck.

Export

An export business (used clothing) was started by a refugee from the Congo who has similar business experience in his country of origin. Although the Vermont Refugee Microenterprise Program assisted the person in creating a business plan and in applying for a loan, he was unable to obtain a loan because the financing organizations considered this a high risk. The Vermont Refugee Microenterprise Program provided the refugee with access to a First Step loan, and he then worked at two jobs for a year to save enough money to get the venture off the ground. By the end of the year, his English had improved so substantially that he was able to work alone with the business counselor. This was in part due to the fact that he attended business planning training and many training workshops. This business generated a profit of 50 percent on the first shipment.

Interpreting services

A refugee from Moldova started an interpreting business. He was able to leverage the IDA Savings Program and a First Step loan. This refugee attended financial literacy and business planning training as well as many business workshops.

Construction

A construction business (subcontracting) was started by a refugee from Moldova.

Other

Through word of mouth, a number of immigrants and refugees who had become U.S. citizens inquired about services. As they were not eligible for the services of the Vermont Refugee Microenterprise Program, they were steered into the permanent Microbusiness Development Program. They received business counseling and several were able to start or expand businesses. These included a taxi business and a furniture retail store that also provided furniture repair and restoration.

Significant findings

These include the following over the life of this program:

- It is essential to have a combination of B/BCs and experienced business counselors to provide interpreting and translation as well as an understanding of ethnic and cultural practices and sensitivities.
- The business counselor provided the explanations of American business practices and processes that are necessary to start and establish a business in the United States. This teamwork was a key ingredient to the success of the program.
- Staffing needs to be flexible to meet the changing needs of the refugee population as new ethnic groups are settled during the life of the program. For example, the program started with Bosnian, Vietnamese, and

Congolese counselors. Later, Russian speakers and Somali (Mai Mai) interpreters and counselors were needed.

- The B/BCs themselves benefited from the learning experiences of the program. In addition to the core team, several interpreters involved throughout the life of the program also benefited from the learning experience. The Vietnamese B/BC is now an eldercare counselor at the Vermont Refugee Resettlement Program, the Bosnian B/BC is employed by the local Red Cross, the Congolese B/BC has completed a master's degree in social services at the University of Vermont, and the Somali-Bantu B/BC is employed by the Burlington Housing Authority as a community liaison officer. The business counselor is now employed by the Vermont Refugee Resettlement Program as the program coordinator for the Vermont Refugee Childcare Training and Employment Program. All these former team members are making a continuing contribution to the community and leveraging what they learned during the life of the program.
- The MicroBusiness Development Program staff in Burlington and other regions greatly increased their knowledge and understanding of the needs of refugees and immigrants through meetings, presentations, and coordination of services.
- Positive publicity for the refugee communities was generated by several well-publicized stories about their business successes. At the very end of the grant, the AP conducted an interview and ran the story in 56 newspapers around the country.
- It was very helpful to have the program under the auspices of the well-established MicroBusiness Development Program, because the Vermont Refugee

Microenterprise Program was able to leverage its resources and programs, such as the IDA Program and the Computer Literacy Program. This also made it easy to transition clients into the mainstream program at the end of the grant period.

- The best forms of outreach were through partners and B/BCs who had connections within the refugee community. Very few clients enrolled as a result of direct publicity efforts. The majority came through word of mouth or were referred by partners.
- Relationships with partners contributed to the success of the program. These partners included the State of Vermont's Refugee Coordinator, the Vermont Refugee Resettlement Program, Central Vermont Community Action Council, the City of Burlington (where the majority of the refugees resided), and nontraditional banks, such as the Opportunities Credit Union.
- Training needs to be flexible and should meet the needs of the community at which it is targeted. With very few exceptions, the majority of the refugees attended training that was specially modified to meet their needs. These needs included: short periods of training in the form of workshops, evening and weekend sessions, and materials delivered in both English and their native languages. Nonliterate clients were a special challenge, and visual materials were developed for these clients.
- Lack of English is a major barrier to starting a business. It was also noted that a significant improvement in English occurred in those refugees who attended the bilingual training and learned business terminology in this way. There was also a significant improvement in those refugees who were able to attend ESOL classes.
- Lack of financing options is a major barrier. The First Step loans provided a start for some of the clients, but a

> number were refused loans due to lack of a credit history (or a poor credit history due to ignorance of credit protocols) or because they did not have sufficient assets.

It's evident that very targeted and tailored programs designed to meet specific needs of the population can be quite effective. Without this program, it is felt that the majority of these businesses would have never been created due to the cultural, social, and economic barriers present for the refugees. By fostering viable businesses in the community, inclusiveness goals are reached, and that's good business for everyone.

Resources and ideas for making it happen in your community

Business Resource Guide for Refugees: www.cedoburlington.org/business/refugee_resource_guide/busref_resguide_toc.htm

Chittenden County Community Action: www.cvoeo.org/htm/MicroBusiness/Microbusiness.html

Mercy Connections: www.mercyconnections.org/

Northern New England Tradeswomen: see article at: www.vermontguides.com/2006/05-may/tradeswomen.html

Vermont Refugee Resettlement Program: http://uscri.refugees.org/site/PageNavigator/Vermont/vermonthome

Vermont Works for Women: www.nnetw.org/about-us

New Americans outreach: Make sure business service providers are reaching out to refugees and new Americans. A large percentage of all new jobs created in the tech sector in the U.S. are in businesses started by people not born in the United States.

Magnet schools: Maintain or create community-based schools. Magnet schools in poor sections of town create good reasons for families to attend these schools. Check out the Sustainability Academy in Burlington Vermont, the first magnet elementary school in the U.S. with a sustainability theme for kindergarten through fifth grades. Visit http://sa.bsdvt.org.

Community in schools: Work with schools to use their facilities after school hours to host community services and events.

Disabled access: Create programs for improving access to buildings and install curb cuts. Twenty percent of the population is disabled.

How to get a business loan: Put yourself in the lender's shoes. If someone was looking to borrow money from you, what would you want to know before making the loan?

4 Crunch and Funk: Cultural Vibrancy

How Is a Culturally Vibrant Community Created where People Want to Live, Work, Play and Visit?

Jerry Greenfield and Ben Cohen, iconic founders of the premium ice cream enterprise Ben and Jerry's, lamented that Burlington isn't as "crunchy and funky" as it was in days past. Recalling the glory days of the 1970s and 1980s, Burlington was a major site for everything alternative, securing a place impression of uniqueness. Everything's relative, however, and Burlington still can hold its own culturally (including being named as the third funkiest city by British Airways in 2005).

What's the deal anyway with culture? A lot, apparently. Study after study shows that culture is a primary element affecting all sorts of outcomes – quality of life, sense of place, ability to attract new citizens and investments – things that help influence long-term community resiliency and economic durability. This chapter provides insight into Burlington's progression towards a culturally vibrant city, the role of a cultural economy sector, and its influence as a factor in contributing to the creation of a more durable economy. It begins with a look at creative city concepts and perspectives, followed by a case in the Closer Look section of the South End Arts + Business Association's transformation of the Pine Street warehouse district into an arts and business incubator district. It's important to note this area was going through disinvestment as a decaying industrial district with warehousing, wholesaling and distribution, as well as some car and truck repair businesses. In essence, it was the city's industrially zoned part of town quickly filling with empty and dilapidated buildings.

Arts, business and culture (ABC)

There are lessons to be learned in how to merge arts and business together. First and foremost, doing so can serve as an effective (and dynamic) economic development strategy, with people and businesses moving in, fostering a creative environment. A popular annual event, the South End Art Hop, provides an example of helping foster a creative environment. This event is a way to showcase a former industrial district, now a working area for arts, entertainment, and yes, still some manufacturing. It's interesting to note that Burlington's major event was first begun by artisans who chose to live in the area, throwing "alley cat" parties that morphed into the formal Art Hop. The event now sees over 500 artists registering each year. This kind of outcome validates who people are, who they want to be, and creates vibrant communities where people want to live – "living values that are valued." The focus is to try to meet the needs of people in the community.

Here's another example: irrespective of political stripes, several citizens got together united by the common goal of making downtown more vibrant. The idea was that if the needs of the people who live here are met others will come here too. Low-income, moderate- and high-income – all will be included. That's one of the ways to make it successful. For downtown, many wanted to make sure families were welcomed, and it was realized that Burlington didn't have enough family attractions. The results? A fountain at the top of Church Street (a veritable magnet for kids), and the ECHO environmental center on the waterfront.

Focusing on sense of place is key – that distinctiveness about a place separating it from others. Some localities "have an attraction which gives a certain indefinable sense of well-being and which we want to return to, time and again."[1] Unfortunately, prevailing development patterns in many areas increasingly separate people from places, reducing responsibility to their communities and threatening sense of place.[2] Burlington isn't one of those places. Its uniqueness isn't by coincidence. It has taken concerted efforts over many decades to create conditions for its uniqueness to unfold (historic preservation, redevelopment efforts and new additions both culturally and to the built environment). The built environment is a reflection of a community's values – what one sees and experiences is directly related to those values.

This definitive sense of place isn't just a strategy to attract tourists, although tourism is a valuable component of the local economy. It's more a lifestyle choice – enabling citizens, businesses and organizations to live the types of lives they want in a vibrant and engaging community. The results are impressive – take a look at the awards garnered in just the last several years. Note that it reflects the cultural dimensions of the city, from historic resources to arts-based activities to a focus on healthy living lifestyles.

Table 4.1 Burlington's awards in recent years

Source: www.gbicvt.org/regional-profiles/burlington-accolades/

June 2009	Burlington's Hill Section named one of ten "Best old house neighborhoods 2009: northeast" by ThisOldHouse.com. The designation was given because the Hill Section's "mid-19th-century housing stock has kept its architectural character and charm," making it one of the "ten best places to buy an old house in the Northeast, from Delaware to Maine."
March 2009	Recognized with the runner-up 2009 City Cultural Diversity Award by the National Black Caucus of Local Elected Officials and National League of Cities. The award seeks to promote and honor community leadership in developing creative and effective programs designed to improve and advance cultural diversity.

November 2008	Designated the "Healthiest city in the USA" by the US Centers for Disease Control and Prevention. Burlington ranked highest in exercise, and among the lowest in obesity, diabetes and other indicators of ill health.
November 2008	Burlington Area deemed one of six "Safe havens in real estate" nationwide according to a review of housing data conducted by Kiplinger.com and published online on Yahoo! Real Estate.
October 2008	The Church Street Marketplace outdoor pedestrian mall designated as one of ten "Great public spaces for 2008" through the American Planning Association's Great Places in America program. The APA "singled out the Church Street Marketplace because of its inclusive and careful planning and design process, historic buildings, thriving retail trade, carefully maintained streets and walkways, and strong community support."
October 2008	Named one of America's "Prettiest towns" in a listing published by the travel website Forbestraveler.com. Burlington was third among the 20 places profiled.
May 2008	Given "Level 5" award (highest possible level) by ICLEI – Local Governments for Sustainability – for efforts to reduce greenhouse gas emissions (one of two ICLEI member cities to receive this award).
March 2008	Named as the "Best walking city" in Vermont and one of the "Top Walking Cities" (ranked #42) in the U.S. by *Prevention Magazine* and the American Podiatric Medical Association.
February/March 2008	Named one of "America's greenest cities" by *Organic Gardening* magazine (ranked second among cities with less than 150,000 residents).

Table 4.1 continued

September 2007	Ranked #1 in a regional study of downtowns commissioned by the *Rues Principales Vieux-Saint-Jean* (compared to 16 other downtowns in Quebec).
August 2007	Ranked among "Best towns" 2007 by *Outside* magazine.
May 2007	Ranked among "Top ten greenest cities" by real estate service, Move.com.
May 2007	"Tree City USA" award by National Arbor Day Foundation for 13th year.
April 2007	Ranked among the top 25 "Small cities and towns" arts destinations by readers of *American Style* magazine.
April 2007	Ranked #5 among "Top ten places to retire young" by *Money* magazine.
March 2007	Burlington metropolitan area ranked as "Greenest city" in the country in survey of 379 metropolitan areas nationwide. Based on air and watershed quality, mass transit use, power use, and number of organic producers and farmers' markets.
February 2007	Ranked second out of 72 cities (50 largest cities in the country plus largest cities in each state) nationwide by the Earth Day Network in its *Urban Environment Report*.
November 2006	Ranked fourth among "Top 10 cities for beer lovers" by ShermansTravels.com.
November 2006	Ranked fifth in "America's healthiest places for women" by *Self* magazine.

July/August 2006	Burlington's Five Sisters neighborhood on the list of "Top cottage communities" of *Cottage Living* magazine.
June 2006	One of the "Top 25 cities for art" in the small cities and towns category – *American Style* magazine.
April 2006	Burlington named the "Best of the best" places to live (top five of their "50 Best Places to Live" by *Men's Journal* magazine.
2006	Among the "Top 10 greenest cities," of homestore.com (AOL Real Estate).
November 2005	One of "50 fabulous gay-friendly places to live" named by author Gregory Kompes.
June 2005	Burlington was number 12 among the top 25 small cities in *American Style* magazine's "Top arts destinations" for 2005.
2005	Ranked as the "Third-funkiest city in the world" by British Airways' magazine *Highlife*.
2005	Economy.com ranked Burlington #1 in the U.S. on its index of business vitality.
2005	Rated one of America's "Dozen distinctive destinations" by the National Trust for Historic Preservation.
2005	"No. 3 best state for healthy kids" – *Child* magazine.

Making places

Many efforts through the years now provide the venues that citizens and visitors alike enjoy. Two long-term residents, Bill Truex, the architect who first conceptualized the project, and Pat Robins, a business owner and planning commissioner, recall the concerted efforts that went into creating Church Street Marketplace, a pedestrian-oriented open mall area in the middle of downtown.

"The genesis of the idea came in the early 70s, when we closed off the street for a fair; later, we built a parking garage to serve the area and closed it permanently," Pat recalls.

> I believe Burlington's focus on creating a market for a public stage has been crucial in its commercial success.
>
> (Bill Truex, originator of the Church Street Marketplace concept)

Many pedestrian malls and similar projects throughout the U.S. have not been successful. Bill and Pat studied other models carefully and came to the conclusion that in order to be successful, there were several key ingredients needed. "We should think of the street as a public space and as a stage for programmed and non-programmed activities, and that this space should be cared for much as a private developer would promote and care for the public space in a private shopping mall," Bill explained. This management of the area is essential, and has been instrumental in its success through the decades. The project was completed and dedicated in 1982. It received a complete refurbishment in 1994, led by Bill and his firm, Truex/Cullins and Partners. Today, Church Street Marketplace is often the most recognized feature of Burlington for visitors, and certainly serves as a hub and gathering spot for locals.

The revitalization of the Burlington waterfront has been an extensive, ongoing project spanning many decades. It took ten years to finally have the oil tanks removed to open up the view from the downtown to the lake and beyond to the Adirondack Mountains. Melinda Moulton and partner Lisa Steele have been instrumental in helping reshape the waterfront into a calling card for Burlington's residents and visitors. "The waterfront has changed from a dump, literally, with ffteen rail lines where no one could access the water at all to a beautiful area," Melinda explains, "this transition wasn't easy, and I believe much of the change for the waterfront and

Burlington overall was prompted by the activist spirit prevalent in the area since the 1960s and 1970s." Jerry Greenfield, co-founder of Ben and Jerry's, agrees, "Talk about access for the public – they made the lake part of the city – it's emblematic of the 'spirit' here."

> *I love the activist mindset and intentional cultivation of the human spirit here.*
>
> Melinda Moulton

Bernie Sanders ran for mayor on the platform that Burlington would have a people-oriented waterfront. Now there are lots of venues there – restaurants, boathouse, ECHO environmental center, bike path, skate park, passive recreation, boating – and it is a destination for festivals. It is inclusive for the entire community. A law from the 1800s public trust document ensures that if waterfront land is filled in, it has to be for public use. So it was protected and codified by zoning. And despite some of the private sector companies in the area wanting alternative development (condominiums, for example), most agree that it is a major resource to have the waterfront for everyone's use. Now, Pomerleau Realty sponsors the annual fireworks display over the harbor.

Since its inception in 1983, CEDO has played a key leadership role obtaining financial assistance, providing technical assistance and long-term planning for the waterfront redevelopment. Early on, the Community Boathouse and Bike Path were developed by CEDO and completed in 1988, providing greater public access to the waterfront. The Boathouse has received an Excellence on the Waterfront Award from the Waterfront Center, and both projects have become Burlington icons. For a variety of reasons, three prior private plans to redevelop the waterfront didn't work, and finally in 1990 the Urban Renewal Plan for the Waterfront Revitalization District was adopted. Along with the 1991 purchase of the Urban Waterfront Reserve from Central Vermont Railway to create Waterfront Park, the modern redevelopment effort was established.

The Leahy Center for Lake Champlain, also known as the ECHO (Ecology, Culture, History, and Opportunity) Center, was initially developed by CEDO; it first opened in 1995 and expanded in 2001. The Center now brings together scientists researching the lake and visitors learning about its ecology, culture, and history. Another project developed by Main Street Landing, the $13.5 million, 62,726 square-foot Lake and College Building, which opened in 2005, is an example of the dense, mixed-use development that has taken place more recently on the waterfront. This project was supported with a $10 million Renewal Community Commercial Revitalization Deduction (CRD) and financial support from the Vermont Downtown Program. CEDO managed the Renewal Community Program, allocated the CRD and worked with Main Street Landing to obtain financial support from the Vermont Downtown Program.

After a number of proposals to revive Burlington's waterfront in the 1970s and 1980s were not fully successful, the Waterfront Revitalization Plan, containing 13 project elements, was approved by voters and begun in 1990.

Figure 4.1 Burlington's fishing pier
Photo: Nick Warner.

Preserving what is valued in the community is essential. With the help of the Preservation Trust Vermont, several key structures have been preserved in Burlington. But it's not about what is preserved, "it's about the context … not just saving a building but putting it to a good use in the community," explains Paul Bruhn, director of the trust. Historic preservation is not just about preservation in Vermont, it's about revitalization of villages and towns, and many historic structures can support these efforts. "A rich cultural and arts community is dependent on tolerance, authentic places, creativity and often these places are historic buildings turned into incubators and affordable space," Paul points out. So it's integrating the built environment into the desired outcomes for a community – in this case, arts and culture are strongly valued. It's pretty clear that the incubator space located in the historic structures along Pine Street, along with revitalization of major focal points such as the Flynn Center for the Performing Arts, have served as catalysts for positive change.

Growing a culturally creative economy

The creative economy is a concept popularized by economist Richard Florida, referring to development strategy emphasizing the attraction of "creative," knowledge-based workers within a culturally vibrant city. Florida compared cities using a metric that examined amenities, social capital, diversity, and several other factors. Burlington scored highly in Florida's index, ranking as the most tolerant and fourth most creative city.

In a 2004 study examining Burlington's creative economy, a number of artistic and cultural attractions meeting Florida's criteria were examined. It cites the example of the Great Harvest Bread Company, which chose Burlington in part over other locations in New England due to arts-based activities and amenities. In explaining why this helped them make their decision co-owner Ethan Brown states,

"The art was just fantastic. Not only did we like seeing the art, but the whole notion of the Art Hop was just cool" (the Great Harvest Bread Company has since become a central feature of this annual event).[3]

This report also calculates the creative economy as playing a quantitatively significant role in Burlington, estimating that between 4 and 12 percent of Burlington's "creative" workforce generate over $500 million annually. Recent reports from Americans for the Arts calculated that 1,040 creative jobs were located in Burlington as of January 2008. The arts district in the South End and Pine Street area contained about a quarter of these jobs, while a little more than half of the jobs were in the downtown central business district.[4] Not to be overlooked is the long gestation period during which this district developed, earning its reputation as "The SoHo of Vermont."[5] Florida makes an appealing argument in correlating certain traits such as the presence of creative workers with economic success, but at what point was the critical mass of creative professionals reached?[6]

First, in the 1960s and 1970s Vermont's reputation as an agrarian and "alternative" place to live attracted many creative risk-takers from out of state, as well as University of Vermont graduates who chose to settle in the state's largest city where they went to school. From the initial *Jobs & People* 1984 report, Burlington has focused on cultivating its economic base and supporting its citizens; however, one concern even in this initial report was the number of underemployed professionals. Incubator space from old manufacturing and warehouse buildings on Pine Street provided an outlet for some entrepreneurs who didn't want to take a conventional employment path or found a dearth of higher level job opportunities.

An influential economic strategy during the 1980s and 1990s was the idea of cluster and network development plans. Forming the

South End Arts + Business Association (SEABA) to focus on development of the creative and arts district, it was initially and explicitly a network, and was gradually transformed into a cluster which has effectively used the creative economy with substantial returns. Like a cluster, SEABA has been able to establish value- and knowledge-adding chains among its members. While this occurs in Burlington on a relatively small scale, SEABA also enables:

> *a scale of demand among employers that produces external economies, i.e., a sufficient number of firms with common or overlapping needs to create or attract more services and resources (including labor) than would be available to more isolated firms – and often at a lower cost.*[7]

However, rather than a mid-skilled labor force with relatively minimal geographic mobility, even in the past this district has more generally housed a "creative economy" workforce.

Also relevant is the comment that "re-orienting the central theme of the cluster from some commonality of production process to a commonality related to knowledge, innovation, or entrepreneurship, may also open up new possibilities for generating externalities and taking collective actions in a region."[8] Rather than companies being the primary player as in a cluster economy, people are the most important resource – and the number of incubators in the area creates a key difference in how the area's economy functions. What are the lessons learned? CEDO's former Assistant Director for Economic Development Bruce Seifer explains:

> *It's about taking your assets and figuring out how to best use them. In this case, it was one building at a time, providing loans, beginning with renovating the ugliest structure in the area, the Maltex building. It was the biggest visual eye sore in the area. We then worked to get the next one, then the next one, until most have been revitalized now. Originally, Bernie Sanders, then Mayor, was blamed for*

> *the vacancies in the area, so we took pictures and one by one we redid them. The idea was to provide places for entrepreneurs to thrive.*

In some clusters, an interlocked labor force can replace social capital in holding such a cluster together. However, in this case, the opposite applies: social capital, and the relationships between businesses holds the area together even considering the vast differences between businesses functions. As the organization and the region has matured, a joint culture has developed allowing for more interlocking functions, particularly among the artists' community, as more individuals are attracted to an area which they feel will benefit them and which they will themselves continue to sustain and support. This development was supported by the initial efforts of SEABA in uniting first to address shared costs but increasingly to promote the district's identity by solving common challenges with infrastructure projects and through more positive community building efforts including the Art Park and other beautification work.

A common concern in arts districts is the issue of gentrification, particularly that fledging companies and artists will not be able to afford rising rents as the area becomes more popular. In the case of Burlington, zoning restrictions have helped prevent gentrification. Language in the city's master plan helps maintain certain types of retail, housing and other district characteristics.[9] Maintaining space for businesses of certain types is vital too – enabling growth of more of the types of businesses that are already here rather than gentrifying for "higher uses."

Let's take a deeper look at SEABA and how this district reflects cultural vibrancy while building solid community economic development outcomes. This chapter's Closer Look case focuses on SEABA, noting its processes and activities over time.

A closer look: arts and redeveloping the Pine Street area[10]

Figure 4.2 SEABA's logo

In large part, the South End Arts + Business Association (SEABA) and the Pine Street revitalization exemplify the concept of a creative economy. Additionally, its relatively recent development allows for a deeper examination of how this "creative economy" originated and the extent to which Florida's model explains its success. But first, some history and context sets the stage.

The Pine Street area has played an important role in Burlington's history, particularly in its connection to the city's industrial base. Manufacturing along the waterfront began soon after the first quarter of the 1800s, and by mid-century the expansion of the railroads into Burlington gave Pine Street properties unique access to a burgeoning transportation network. During the 1870s, much of the remaining swampland in the area was filled, and the additional space was used to develop several manufacturing buildings. However, the railroads that connected Vermont to Canada, New York, and Massachusetts would also eventually allow companies from more distant locations to undercut Burlington merchants. By the 1890s, the lumber industry was in decline, having encountered tariff changes, competition from western timber, and several financial panics. However, "the street's easy access to the railroad, available factory space, and ready supply of labor in the King Street and South End neighborhoods, made the transition to new industries relatively simple."[11] At the turn of the twentieth century,

> *the Maltex Building housed the Malt Food Company, cereal manufacturers later renamed Maltex, the inventors of Maypo; E. B. & A. C. Whiting Company commercially dressed the fibers for brooms and brushes; and the Welsh Brothers Maple Company blended pure maple with cane sugar, developing Vermont Maid Syrup. The Queen City Cotton Company built a factory on Lakeside Avenue, employing over 600 workers. A light and temperature controls factory on Flynn Avenue was built by Lumiere North American Company.*[12]

These newer industries didn't rely on lake transportation, and they would soon be supplemented by transportation and distributing companies as the automobile became more ubiquitous.

> *St. Johnsbury Trucking built its large modern facility in 1941 ... Its location between the center city and suburban areas however, [made] it an ideal spot for the repair shop or tire store that needed ready auto access, but not the high visibility or expensive land of the Shelburne Road and Williston Road strips.*[13]

However, as the twentieth century progressed, many of the larger factories either closed or downsized, and while some buildings were subdivided to make room for smaller-scale industrial uses, others remained vacant. The only exception is Blodgett Ovens, started in 1848 and still in operation, producing commercial kitchen ovens.

The roots of Pine Street's and the South End's current business revival began in the second half of the twentieth century. A local developer, Ray Unsworth, had inherited the Whiting Company from his father in 1950. (Ray's father, Thomas Unsworth, had moved to Burlington in 1913, merging his Brooklyn-based Brush Fiber Supply Company with the Whiting Company.) The Unsworth family sold the business in the 1950s, which moved to another Pine Street location in the 1960s, leaving the original cluster of seven buildings to become vacant.[14] In the 1970s, Unsworth saw the original factory space deteriorating and decided to transform the site, known as the Howard Space Center, into a number of incubator buildings. The concept of "incubator space" was still new and Unsworth was likely unaware of the term[15]; however, his work was recognized in a *Vermont Life* Magazine story in 1982,[16] and "other entrepreneurs follow[ed] suit, providing artisans and small businesses with affordable, more professional environments in which to do business." This environment would set the stage for further public and private efforts in the 1980s to redevelop the remaining unused buildings and form a business network for new and older businesses (and later artists) in the area.

Getting organized

CEDO looked for innovative strategies and creative funding mechanisms to achieve its mission, and the organization soon saw opportunity in the Pine Street area. One of CEDO's first economic development strategies from its 1984 *Jobs & People* publication, which provided a framework for CEDO's economic goals, was to promote a diverse base of locally and employee-owned companies within Burlington.[17] Given the incubator space and small businesses already beginning to grow in the area, Pine Street appeared poised for a revitalization attempt that could become a centerpiece of Burlington's vision for its future economy. CEDO used two primary strategies for the Pine Street area. First, CEDO worked with other public and private groups to continue the work Unsworth had begun in the 1970s of redeveloping Pine Street buildings. These specific efforts will be discussed in the next section. Second, working with state legislator Bill Mares, CEDO helped to conceptualize a business association for Pine Street companies. After suggesting a Pine Street/South End fair for businesses, CEDO realized that the businesses would need to become more organized for such an event to take place. Looking into how such an organization might operate, CEDO sent a survey to Pine Street companies and began to meet with interested parties. CEDO had coordinated workshops and classes, including one specifically for Pine Street incubator properties, which organizers felt could be incorporated into the new association.[18] A similar group with 60–70 businesses had also recently been developed in central Vermont, which could serve as a potential model for the group. SEABA formed as this new group.

The explicit purpose and goal of SEABA focused on saving money and assisting businesses. Initial proposals focused on cost-saving tools, such as discounts on health insurance or joint marketing as well as some emphasis on the development of Pine Street infrastructure, including better signage and road repaving. Bruce Seifer of CEDO specified at the first meeting that:

> *the City is interested in assisting to get this organization off the ground, but they would not be there in the long term to direct this organization. Instead, they would like to help it be developed and then slowly wean itself away from this organization.*[19]

This allowed businesses who might not always support the city's actions to join, and gave companies a feeling of ownership. This was particularly important given a certain level of political tension between business owners and the City following Bernie Sanders' election as mayor.

Original SEABA Statement of Purpose, 1986

The Greater Pine Street Business Association (GPSBA) was formed for numerous and diverse reasons; however the primary goal of the association is to benefit businesses in and around Pine Street. The association will have a broad and ever changing role in enhancing and serving individual Pine Street businesses and Pine Street as an entity. The association will act as a force uniting the positive energies, and emphasizing and addressing problems concerning Pine Street businesses. Thus enabling Pine Street to mature toward being a more attractive and economically recognized area to conduct a diverse mix of businesses providing products and services.

The GPSBA will attain this prosperity in a number of ways. The association will act as a spokesperson for Pine Street, representing a recognized large group of constituents, lobby local and state authorities and public issues. The association will provide business literature, seminars and workshops to educate area businesses concerning advertising, marketing, insurance, accounting and other business skills and needed information. The association will also provide a network of local services or expertise which Pine Street businesses can utilize within the association. The association will also provide a bi-monthly Newsletter recognizing Pine Street business activity, resources, and happenings.

Following Unsworth's efforts in the 1970s, CEDO's formation in the 1980s allowed the City of Burlington to become more involved in Pine Street's development. CEDO provided assistance to developers as well as actively seeking out an Urban Development Action Grant (UDAG) to help restore the Maltex Building. SEABA's role in developing and demonstrating Pine Street's growing vitality was also important as a signal to other private investors, as exemplified in the redevelopment of the Kilburn and Gates Building in 1988. The following bullets list projects which were supported by CEDO prior to 1990 as described in an application for the 1990 James C. Howland Award for Urban Enrichment by the National League of Cities, which granted an honorable mention to the Pine Street Revitalization Project.

- *Restoration of the Maltex building at 431 Pine Street*: This vacant, 40,000-square-foot building was converted with an Urban Development Action Grant (UDAG) in 1985 into an "incubator" for small businesses to start-up or expand into. After completion of the project, it became home to 20 businesses employing 105 people (1990). This building is significant in that it was the first major building to be rehabilitated on Pine Street.
- *Pine Square complex at 266 Pine Street (now called the Soda Plant)*: The 40,000-square-foot "old Coke plant" was bought and renovated in 1985. CEDO helped obtain the industrial revenue bond for this to happen. Over the next five years (1985–1990), five of the businesses in this building were expanded and moved in to buildings of their own. In 1990, there were 22 businesses operating out of the Pine Square complex.
- *The Vermont Maid Maple Syrup building at Pine Street and Marble Avenue*: Gaslight Home-style Laundry Center obtained a $50,000 low-interest loan from CEDO to renovate a portion of this vacant building. After the laundry opened, ten new businesses moved into the building. The laundry service is the largest in Vermont and provides a needed service to the Pine Street area neighborhood residents.

- *The Kilburn and Gates building at Pine and Kilburn streets*: In 1988, a 100,000-square-foot vacant building was redeveloped by the private sector as a result of the successful incubator developments on Pine Street. In 1990, 29 businesses had leased space in this building, including the area's first health clinic.

Strategies: incubators, marketing, and networking

Incubator buildings, such as those described in the previous section, played a key role in CEDO's and Burlington's economic development strategy of local ownership and the fusing of culture and commerce. Incubators offer shared services to start-up or young companies in a supportive environment, often at favorable costs. Rather than only trying to attract larger businesses, CEDO theorized that working to grow smaller businesses would result in a more balanced pattern of economic growth. First, local owners would ensure more stable employment opportunities due to a deeper understanding and interest in Burlington's economy in the long run. Profits and wages would be more likely to be reinvested within the community, and locally-owned businesses would be more likely to hire and train local residents. Particularly in the case of incubators, owners who were successful and expanded would also be more likely to remain in Burlington and give back to the community.

There are many arts and business networks throughout the United States and around the world. These associations were typically created to provide business resources for artists and other creative professionals. There is often much interaction, with local business leaders serving on the boards of nonprofit arts and cultural organizations. However, the level of integration SEABA has provided is less common, as there is usually less exploration of how artists can support and promote local business. Many of the more prominent organizations are located in larger metropolitan areas such as Philadelphia, New York, and Chicago, although others do exist in smaller communities.[20] Another element of the local ownership strategy was a "Buy Burlington" campaign which soon developed a Pine Street equivalent, "Get it on Pine Street." This strategy had begun before SEABA was formed and was mentioned in the minutes at the initial meeting; however, the organization was able to more effectively advertise the slogan and begin a city-wide process of rebranding Pine Street.[21] SEABA also established a Pine Street Discount as a way for businesses to support each other while improving their own sales. In more recent years, companies like Burton Snowboards, whose world headquarters are located in the South End, have continued this tradition by contracting with local creative firms such as Select Design, RL Photo Studios, and JDK. Various directories, including a current online version, have also listed SEABA's members, promoting Pine Street businesses throughout Burlington and the region.

Incubators have traditionally been an important resource for start-ups and other small businesses. While companies on Pine Street offer a wide range of products and services, incubators can serve as forums to provide business guidance as well as social and moral support for many of the common challenges facing young businesses. SEABA's existence expanded this service, connecting small businesses with medium and larger-sized companies. There are several ways that SEABA fosters connections to a wide range of community resources, including hosting street parties and offering workshops and counseling to entrepreneurs.[22]

These efforts have continued to the present, with resources available for South End businesses as well as networking opportunities at events such as the annual meeting or Flamingo Fling fundraiser. This community that SEABA has helped foster serves an instrumental function for the district, promoting a collective identity. This in turn attracts others who want to be a part of the district and its culture.

Early initiatives

Health insurance

An important early SEABA initiative with present-day implications was the network's attempt to provide health insurance for its members. Initially, an obvious benefit of SEABA was the opportunity for collective savings on major costs such as health insurance, which was prohibitively expensive for many small businesses. Even in the late 1980s, over half of small business owners in Vermont and many Pine Street companies were unable to provide this coverage. By purchasing health care collectively, SEABA hoped to negotiate better rates with insurance companies so that employers could then offer better coverage to their employees.

The process of offering health insurance began relatively early on in the organization's history. In November of 1987, SEABA sent out a survey to members and other Pine Street businesses, inquiring whether they offered health care to their employees and whether they would be interested in learning more about a group policy. SEABA found a number of challenges in selecting a group policy. These included differences in the types and risk levels of businesses, varying health insurance needs which differed by age and marital/dependent status, and a lack of administrative support.[23] At this time, that there were no staff members for the organization added to the complexity of the situation. The insurance subcommittee was able to find a plan which offered a volume discount, but not a complete group rate. SEABA selected the Washington National Classic Care Plan by Mutual of New York in August of 1988.[24] In early 1989, however, the company temporarily suspended granting new policies in Vermont, and this decision was made permanent in May 1989. Pine Street businesses were still able to cover employees who already had insurance, but they were unable to offer policies for new employees. Fourteen companies had by this time obtained policies for their employees. A new committee was formed to explore further insurance possibilities, but by August of 1989, the difficulty of finding a group policy and lack of interest led the organization to decide not to recommend a plan. SEABA did try to raise awareness of some local resources for companies looking to provide health insurance to their employees, but group health insurance no longer appeared feasible.[25]

Daycare

In contrast, a more successful early SEABA initiative examined different proposals to reduce childcare expenses for Pine Street workers. A questionnaire was developed in the summer of 1987 to determine the needs and interest levels of local workers. As a result of this survey, a daycare center operated by Baird opened in November, 1988. This center was originally licensed for 15 infants and toddlers aged 6 weeks to 3 years old; however, by July, 1989, it expanded to offer childcare for 35 children aged 6 weeks to 5 years old.

Similar programs were also being developed at that time; a collaborative effort in Portland, Maine for small- and medium-sized businesses working with a university was just one example of how such a daycare center could be established and likely provided at least some inspiration for SEABA's program.[26] SEABA also enabled Pine Street to have a voice on how daycare costs were affecting them. Members actively participated in the mayor's and governor's task forces on daycare, and there were several opinion pieces in Pine Street newsletters advocating for more support for both families and childcare workers. Even today, at least one daycare center, the Pine Forest Children's Center, a nonprofit organization formerly operated by Baird, is still in operation in the Pine Street area.

While the group health insurance option was full of possibilities, the challenges of establishing and implementing the program proved too difficult to outweigh the relatively limited benefits given the difficulty of obtaining a group rate. The experience also implied that it would take significant effort for this initiative to be modeled in other communities. SEABA itself was frequently in communication with state officials in an attempt to select the best policy, and found difficulties when new officials came in who had less information about insurance options. Overall, this project illustrates a primary challenge an organization such as SEABA faces; given its relatively small size, there simply isn't the administrative capacity to effectively reduce costs for such a complex expense.

However, a childcare solution was much more feasible. Providing support and connections for an external company with a stake in the area offered a higher probability of success than attempting to run a daycare center. This model also invites similarities to CEDO and the greater Burlington economy, which has historically used nonprofit and for-profit companies as well as networks like SEABA to support and strengthen infrastructure for social enterprise. In both cases, surveys were useful in measuring the interest in a given proposal; however, they couldn't always measure the extent of the contribution that companies were willing to make. Both projects also demonstrate the need of a core group of dedicated members who are willing to take a leadership role, as most members weren't as highly involved even if they appreciated the benefits insurance or daycare might offer.

Renovations, beautification and infrastructure projects

In 1989 and 1990, SEABA applied for two Community Development Block Grants to enable aesthetic improvements to the Pine Street area. This was accomplished by working through the local

Neighborhood Planning Assembly, one of a set of councils to which the city delegated responsibility for awarding funds for neighborhood improvement projects. SEABA worked with neighborhood residents and the Department of Parks and Recreation to improve the landscape, planting bulbs, wildflowers, and trees. SEABA also helped advocate for improvements to the street itself, leading the Public Works Department to repave the street and install sidewalks. These early developments also helped to improve the image of Pine Street and better integrate it with neighborhood residents.

In later years, the Five Sisters Email Forum, an email listserve for neighbors to stay connected, provided another way for SEABA to communicate with residents who live in the South End.[27] This project would eventually spread throughout the city (and beyond) as Front Porch Forum, and is illustrative of the web of connections that SEABA is a part of which build different levels of community in Burlington and allow citizens to communicate and advocate for their common interests. CEDO provided technical support to get the Front Porch Forum started; now they're going statewide with Knight Foundation funding.

Other infrastructure projects helped shaped the area too, including those relating to transportation, the Southern Connector (Champlain Parkway) and the Barge Canal, as well as environmentally related projects such as Engelsby Brook. These are described as follows.

The Southern Connector (now called Champlain Parkway)

The Southern Connector was initially proposed in the 1960s by City and state transportation authorities as a limited-access highway to reduce traffic congestion for individuals commuting in and out of Burlington. Initially termed the Burlington Beltline Project, the goal was a four-lane highway which would provide a North–South freeway through the city. The northern section of the Beltline was completed in 1971, connecting the North End with downtown, but right-of-way acquisition on the lakefront, residential displacement, and lack of funding led the City to reject the central city portion of the highway. However, there was still support for improved access to the downtown from the south, and the project was reformulated in 1975 as the Southern Connector. This proposal involved constructing several sections of highway (termed C-1, C-2, and C-8) which would connect the I-189 interstate with Battery Street through a newly widened and extended Pine Street.[28]

The emergence of the Pine Street Barge Canal as an environmental hazard and its designation as a Superfund Site in 1981 meant that this initial plan would need to be modified. This led the state and City to consider temporary alternatives. (The four-lane C-1 section closest to the interstate had been built in the late 1980s but never opened.) Several alternatives were evaluated throughout the 1990s, and in 2002, after extensive public comment, the modified 1997 Selected Interim Alternative was chosen. This option (C-6) routed traffic along Pine Street and Lakeside Avenue rather than constructing a new roadway. During this time period, the Southern Connector was also beginning to be referred to as the Champlain Parkway to distinguish the planned residential two-lane road from its initial four-lane highway status. Today, the city is still working with state and federal administrators to build the Parkway connecting Pine Street directly with Main Street.

While the Southern Connector project began before SEABA was established, the organization soon demonstrated its ability to speak on behalf of South End businesses. SEABA sponsored a mayoral debate in 1987, which asked mayoral candidates Bernie Sanders and Paul Lafayette a number of questions related to Pine Street, including one regarding the Southern Connector.[29] During the early 1990s, SEABA was influential in supporting an alternative which would keep traffic on Pine Street while avoiding an increase in the number of lanes. SEABA also provided a seat at the table for a member to participate in a "Sounding Board" commission on the project.[30]

SEABA also sponsored a forum to consider what Pine Street should look like in the future. A SEABA letter opposing the original Southern Connector plan emphasized the necessity to "designate/ preserve the character of the community" and the Pine Street identity, as well as expressing concern about "protecting the economic vitality" of local businesses and the area's industrial past.[31] In 1992, the organization hosted a "Redesign Pine Street" night with community members to discuss suggestions for how the Southern Connector project could incorporate neighborhood improvements such as better signage and sidewalks, bike path additions, and landscaping.[32]

Despite having a seat at the table and hosting numerous public meetings, SEABA still struggled to have the City incorporate its recommendations into the plans for the Southern Connector. The process wasn't particularly efficient: while SEABA was able to help prevent a limited-access highway from damaging local businesses, there were few positive transformations. The aesthetic and general landscaping improvements typically only occurred when SEABA took the initiative to complete them, as was described in the previous section. Ultimately, more than 40 years after it was initially proposed, the Southern Connector has still not been built.

These projects during the early 1990s formed a second stage of SEABA. The organization continued primarily because there was a use for the collective resources of SEABA to address a common challenge which most business owners wouldn't be able to effectively address on their own. Without this common threat, and with the major cost-saving opportunities exhausted and properties generally developed, SEABA would have likely taken a back burner to other community issues; however, this did allow SEABA to expand its reach and become better known in the South End and citywide. The common defense of the Pine Street identity also set in motion the next stage, as beautification efforts transformed into interest in an art park, and increased emphasis on the arts community as Pine Street became increasingly part of the "creative economy."

The Barge Canal

The Barge Canal was dug in the nineteenth century to serve as a shipping access between Lake Champlain and Burlington's waterfront industries. From 1908 to 1966, however, the canal and nearby wetlands were contaminated by coal tar, cyanide, fuel oil, and other contaminants, byproducts of Burlington Gas Works' coal gasification plant which were dumped into the canal. The canal began to occasionally leak oil into the lake, leading the Coast Guard to block the canal from the lake and ultimately for the site to become Vermont's only Superfund site in 1981. Plans for the Southern Connector had initially called for the route to pass through the Barge Canal, and as a result, both the City and state either owned or had some level of responsibility for the site. This meant that both were ultimately named as EPA "Potentially Responsible Parties" (PRP) and might be held responsible for some portion of the final costs to clean up the site.

After several million dollars were spent on studies during the 1980s, the EPA decided on a plan in 1992. The EPA report stated that:

> *The preferred cleanup alternative calls for the construction of a containment/disposal facility (CDF) over the most heavily contaminated portion of the Site, principally the wetland area west of the former coal gasification plant where subsurface free phase contamination is located. The CDF area will be made by constructing subsurface vertical barriers and will ultimately be covered by a low permeability cap.*[33]

However, there was considerable opposition to the cost of the $50 million plan. There were questions about the validity of some of the EPA's findings, such as whether they had overestimated the dangers of the site in its current form or had underestimated the potential health hazards of dredging up the chemicals. A grassroots opposition organized and began publicizing its views. One newsletter argued:

> *First, a wetland will be destroyed to create a landfill. Second, as the toxic waste is removed from the Barge Canal, aromatic toxins will be released in to the air, undoubtedly endangering human health in the area. Then, when the waste is placed in to the landfill its weight might squeeze toxins out of the spongy marsh into the lake. And after all these efforts, with their attendant health risks, only about 25 percent of the contaminated soil will be landfilled; the rest will remain right where they are now. Finally, a 25-foot structure bigger than the South Burlington University Mall will be left sitting on the shores of Lake Champlain for future generations to look at and worry about and continually maintain.*[34]

SEABA was active in these efforts to oppose the EPA plan. In a letter to the EPA, SEABA supported a solution which would allow the land to remain usable, even though the companies which had contaminated the site would likely want to choose the least expensive solution of capping the area. Members attended public meetings held by the EPA and distributed leaflets opposing the EPA plan. These efforts demonstrated the collective power SEABA had achieved in its ability to inform the administrators and members of the public of their position on the issue.

Ultimately, the EPA withdrew its initial plan in 1993, allowing a "Coordinating Council" to try to discover a "Vermont solution"[35] that required consensus among all participants. This group included the EPA, PRPs, and community members, including SEABA board member Marty Feldman. This became a model for the EPA on how to deal with contamination and get community involvement in finding solutions. Marty represented the business community. The group spent the next three years collecting more information, negotiating and compromising over the extent of information necessary versus its cost. Ultimately, after 2,000 hours and four and a half years,

> *Consensus was reached and the Record of Decision was finally drafted. The remedy was a sub-aqueous sand cap on the canal sediments and a weir, or underwater fence, between the canal and the lake. The cost was under $4 million. This number, plus an additional $4 million to the EPA for past studies was dramatically less than the original $50 million remedy.*[36]

Given the drastic difference in the estimated cost of remediation, Feldman and the Lake Champlain Committee, another local community group, felt that there was an opportunity to restore the original wetland loss from the project:

> *A $3.5 million, three-part project, paid by the PRP's would accomplish: a lake ecology research center, run by UVM; a storm water cleanup of Engelsby Ravine, which was a collecting point for urban runoff into the lake about one half mile south of the barge canal; and funding for the city to administer a Brownfields project to upgrade building sites that were economically damaged by environmental harm.*[37]

The other business PRPs were resistant to this proposal, but the group discussed the proposal with local and state politicians, obtaining support from Senator Patrick Leahy and Representative Jim Jeffords. Finally, proponents gained the backing of the Green Mountain Power PRP representative, and after extensive discussion "The Special Projects" were adopted.

The final compromise solution was decided in 1998, to be enacted in 2000.[38] The solution was a model case where consensus-based decision making was successful, but the level of commitment required cannot be overstated. The Barge Canal had demonstrated SEABA's power to speak on behalf of Pine Street and the South End, but it also showcased the necessity of motivated individuals who were willing to take on responsibility for objectives that were important to the group. Without Feldman's efforts, the additional environmental remediation projects likely would not have happened.

The Engelsby Brook project

The Engelsby Brook Watershed is an approximately one square mile area which was degraded by substantial stormwater runoff and phosphorus pollution, forcing the long-term closure of the local Blanchard Beach in 1992. Advocacy came from Independent/ Progressive City Counselor Mark Kornbluh, who was elected in part to the Ward 5 City Council seat with the platform to clean up and reopen Blanchard Beach. After several studies, in 1999 the City of Burlington developed a comprehensive $1.8 million restoration plan (using partial funding from the Pine Street Barge Canal Settlement, and other EPA funding).[39] The plan focused on "residential property owners, school based programs and general public educational efforts, such as those supported by the Chittenden County Regional Stormwater Education Project."[40]

However, there was little effort being made to support pollution prevention among businesses, whose properties totaled 23 percent of the watershed area. This led to the development of:

> *The Business Friends of Engelsby Brook project, a collaboration of Lake Champlain Sea Grant and Friends of Burlington Gardens, [which] informed non-residential property owners/managers in the watershed about water*

quality issues and pollution prevention, and provided specific guidance and technical support to promote adoption of lower input landscape maintenance practices.[41]

Eighteen of thirty-five contacted properties agreed to take part in a survey on landscape practices, and ten agreed to take part in the program. "These 10 owners/managers, including two contracted lawn care companies, managed 16 of the priority properties," which "accounted for more than 50% of the total area of priority business properties in the watershed."[42] Their efforts achieved "an estimated reduction of 0.45–0.93 metric tons of phosphorous," which was substantial considering "the entire Lake Champlain basin phosphorus reduction target is 80 metric tons annually."[43] A similar project was planned for Rutland in 2006, and there was also interest in replicating the program across New England as a way to help businesses save costs and promote pollution prevention.

SEABA's primary involvement in this project was through its participation in the Pine Street Barge Canal Settlement, which obtained the initial funding for the project. However, it is likely that the community and social responsibility which it helped to foster among South End businesses was likely influential in the way that companies were willing to come together and take action to solve a collective problem. SEABA did play a role in other aspects of the cleanup process, organizing two volunteer efforts and several dozen volunteers to clean up the Engelsby Ravine in the spring of 2000. Some of the trash that was collected was even displayed as "object art" at a Pine Street art gallery, Flynn Dog.[44] Fifteen years after it had closed, Blanchard Beach reopened on June 27, 2007.

All about artists

A continuation of the beautification project begun in 1989, the Maple Works Art Park was discussed as a possibility for the next several years until serious efforts began to see if it could happen.

SEABA's executive director talked to local developers, the Unsworth family, who agreed to donate the land, and worked with local architect David Coleman, who agreed to design the park. SEABA then worked to raise money from the local Ward 5 NPA as well as from other fundraising efforts, which ultimately included $15,000 from various sources.[45]

> *On the corner of Marble Avenue and Pine Street, in front of the historic Vermont Maid Maple Syrup Building, is the site of the Art Park developed in 1992 through the efforts of the Pine Street Arts + Business Association. The purpose of this park was to have an attractive and restful public space on Pine Street in which to view a display of nature and outdoor art. Up 'til then there were very few trees and flowers on Pine Street.*[46]

The plan was also to familiarize passers-by with the offerings on Pine Street with a map and listing of artists and businesses in the area. This directory included information about the history of the

Figure 4.3 The Soda Plant, on Pine Street, houses many artists' studios which are open to the public
Photo: Bruce Seifer.

Pine Street area, and there was also information on how to make area businesses more energy efficient.

While the Art Park was successful for the next several years, by 1994, artworks were no longer being regularly rotated in the park. An effort to revitalize the park was spearheaded by Lorna-Kay Peal, the SEABA executive director at the time. A SEABA committee was formed which submitted a request for proposals to the community. Ultimately, landscape architect Ken Mills was chosen to rebuild the park. Many companies and community members supported this effort with donated time, money and materials, and today the Art Park is maintained by the Burlington Garden Club.

For the Art Park to be successful required a major collaborative task, with substantial time and effort put in by a few people, over a long period of time. It is not surprising that the motivation to keep the park running as well as it had been in the beginning declined after several years. However, that the Art Park was revitalized after several years was a positive sign. Even in its first lifecycle, the park created a space which helped to provide a sense of identity for the Pine Street community beyond its traditional industrial past, establishing a sense of place and proposing that Pine Street might soon become a destination.

A lesson learned? Industrial parks are not really parks, since there are no sidewalks and they are not designed for interaction. Rather, this area has been transformed into creative spaces for interacting with each other and art, including coffee shops housed within a business incubator building ,or small pocket parks for having lunch or outdoor meetings. Art Hop is a wonderful example of this as well – by providing exhibitions throughout the district, art becomes a part of the space. With sculpture up and down the streets, including a half-moon shopping carts conversation piece, there is art in every business, helping create a different mindset and spurring innovation and creative thinking. It's intentional – by supporting innovation in businesses and the community, arts will support commerce. Using sculpture in an industrial area helps fuse the idea of commerce and culture.

The Art Hop

What has now become the defining feature of SEABA is the annual Art Hop, an event which first started in 1993 with 36 artists in 27 locations and continues to the present. The first Art Hop was actually an independent production coordinated by two Burlington residents, both of whom visited Columbus, Ohio's event to learn how to make it happen in Burlington. SEABA took over the Art Hop in 1994, gradually increasing the number of attendees from fewer than 300 to more than 40,000 visitors during the month of September.[47]

From the beginning, the Art Hop has been devised as a way to showcase artists and connect them with the community by displaying their works in local companies and venues throughout the district. It promotes exposure for both the host companies and the artists, and is open for all to display their creative works. Even some of the businesses get in on the act, with employees displaying their works of art. It provides the opportunity to explore the businesses in a fun environment, opening up visitors to the beauty and power of creativity. The idea is not solely focused on marketing for tourism, but rather for locating business there, and marketing to families to make an industrial area more human-scale in character. All of this serves to focus on defining interactions of arts and business. The area has been branded by creating a space and a mindset that this is where it all goes on for arts and business. The Art Hop has always been an organizational challenge for a nonprofit dependent on a sole executive director and interns to coordinate the event, as well as to obtain funding from a wide array of local sponsors. However, the Art Hop has also brought a good deal of awareness of the artistic community in Burlington and across the state, receiving an award as one of Vermont's "Top ten fall events" by the Vermont Chamber of Commerce beginning in 2002. The Art Hop has also had direct economic benefits for the area in attracting new talent to Pine Street and the South End, which will be discussed in the next section.

An update from SEABA's director

In the years following the work on the Barge Canal, and the Southern Connector/Champlain Parkway, SEABA continued to present the South End Art Hop, and the event steadily grew in popularity, size, and scope. So much so, that the administrative activities of the organization, and the popular understanding of SEABA's profile, were both overwhelmingly linked to the event.

SEABA's executive director from 2009 to 2011, Roy Feldman, saw it necessary to reinvigorate the broader mission of the organization.[48] Many South End businesses expressed the need for better municipal promotion of their creative and commercial activities, now that the nature and complexion of the South End had changed significantly in the decades since the once purely industrial and manufacturing activities of the district had evolved into an area of creative production. Those involved in the creative economy in the South End needed to be fortified by a means of directing residents and visitors to their activities. Area businesses and artists also sought to be acknowledged for helping develop this profoundly unique district of the Burlington and regional community.

Feldman worked in ensemble with city councilors from three different municipal wards in constructing a resolution calling for the establishment of the South End Arts District.[49] The most critical aspect of establishing the South End Arts District included the installation of signs to attract and direct interested parties to the various galleries and studios.

To accommodate visitors, SEABA's offices were moved from its hard-to-find location at 180 Flynn Avenue, to the center of the district: 404 Pine Street, directly across from the Maltex building. SEABA sought and received designation as a State of Vermont Visitor's Center, and as a visitor center will be directing individuals to the many attractions within the district, including a coordinated system of open studios, scheduled on a year-round basis. Visitors will be directed to visit the artist at work, and in doing so be given the opportunity to provide direct support to Burlington's creative economy.

SEABA's other recent accomplishments include: expanding revenue by 35 percent, including a 200 percent increase in grant awards; reviving membership beyond South End Art Hop registration, and increasing membership benefits; dramatically increasing membership activities and events, from twice a year to monthly; working together in collaboration with other area arts and business organizations; and increasing the size of the board of directors to its potential 12 members and reactivating a committee system.

There is great promise for the future of SEABA, and through the organization's efforts, a benefit for the entire community.

Beyond Pine Street

SEABA's influence and inspiration is felt beyond its district. The creation of the Old North End Arts and Business Network in another area of Burlington provides an opportunity to examine the role of SEABA and its effectiveness. SEABA's success does not necessarily imply that it could easily be replicated in other locations, but there are certainly elements which could likely help other communities.

There are several distinct differences between the two organizations in both their mission and community demographics. Pine Street was traditionally an industrial corridor to the lake, with relatively little direct residential focus, unlike the Old North End. While both locations offer substantially lower commercial rental rates than in the downtown, the Old North End simply doesn't have the scale to replicate Pine Street's industrial environment. However, some efforts have had moderate success in building a certain level of community. The Old North End Ramble has brought individuals from all over Burlington to walk through the Old North End along North Street and gain familiarity with the shops and services offered. The First Friday Art Walk, which includes a number of stops in the Old North End, attempts to provide the same exposure to artists in the area. The First Friday Art Walk was in fact conceived of by SEABA as a logical extension of the Art Hop, and as a way to unite art in the Old North End, downtown, and South End. Finally, a farmers' market provides better connections within the neighborhood as well as some heterogeneity by geographical region. The Old North End Arts and Business Network website provides the information, perhaps the best indication of the belief that creativity can be a basis for encouraging regeneration:

> *Our Mission: To stimulate the economic vitality and enhance the diverse mix of Burlington's Old North End arts & business community.*
>
> *We feel a strong connection to the North End and are taking efforts to promote our unique blend of art, commerce, and entrepreneurial spirit. We recognize the importance of the North End as a presence in Burlington, and are taking efforts to support the vitality and sustainability of our community. To fulfill this mission several objectives have been identified that we adhere to and provide for the North End community. These include:*
>
> - *promoting voluntarism*
> - *building arts management capacity*
> - *gathering and disseminating pertinent information*

- *advocating for closer ties between business and the arts*
- *informing our members of issues that affect our community and actions we take on its behalf.*

We want our organization to become an influential voice for our members and the community; to address, support and ensure their common interests. Through our actions and initiatives we hope to transform the public perception of the Old North End. In order to be successful, ONE Arts & Business is taking advantage of mentoring opportunities with a number of existing arts and business organizations. The City of Burlington's Community and Economic Development Office (CEDO), the South End Arts + Business Association (SEABA), and the Burlington Business Association (BBA) are just a few of the partners that we have been working with. Joining forces with existing organizations is a natural fit to capitalize on the synergies between all areas of Burlington. Utilizing the existing resources of various organizations will help launch our fledgling project, and create a stronger relationship between the Old North End and the rest of the Burlington business and art community.

(ONE Arts and Business website, www.oneabiz.org/about.html)

Overseen by CEDO, the North Street Revitalization Project, with a final cost of $6.6 million, improved streets, lighting, and sidewalks; added a community art center; and partnered with non-profit foundations and private architects to offer Facade Improvement programs for commercial buildings within the project area.

Summary

Art and culture can play a powerful role in community development. By incorporating art into all aspects of this industrial district, a venue for social interaction has been created as well as building capacity to support long-term effort.[50] SEABA began with 30 businesses and artists; today its membership totals in the hundreds. The organization has sustained itself and successfully aided a transition to a new "creative economy." SEABA has been strongest when it is issue-focused, although this has presented some challenges to how others in the community perceive it. Currently many people have difficulty conceiving of SEABA as distinct from the annual Art Hop.

The organization has also generally been highly dependent on the efforts of a few key participants.

Funding has been relatively minimal and dependent on membership dues and donations from board members and local businesses. Some grants were applied for and occasionally received during its history. The organization has depended heavily on unpaid interns as well. The executive director position was initially only 8–12 hours per week, although this has grown to be full-time.

To remain relevant, SEABA continues to focus on the needs of Pine Street and South End companies, relying on its ability to recognize and coordinate services for businesses rather than necessarily providing them itself. For example, SEABA will continue and increase its involvement with efforts to design the Champlain Parkway, advocating a balanced approach that reduces traffic congestion while also supporting street amenities. Improvements to sidewalks, curbs, and trees, as well as the addition of several pocket parks with benches or illuminated sculpture sites, would contribute to maintaining the Pine Street culture and identity. Additionally, directional signage would provide a substantial benefit to Pine Street in designating and promoting the South End as a business district. Hairpin signs were originally developed for the City Department of Public Works over 20 years ago, but plans to install them never materialized.

SEABA must also continue its successful marketing and branding of the South End by sustaining events such as the Art Hop, while ensuring that the event doesn't detract from the organization's original purpose as a resource for area businesses and artists. Future challenges will depend on what the mission of the organization will become. SEABA may become more important if there is a collective need on behalf of the South End businesses, otherwise it may remain relatively low key, but most important is its ability to fulfill its purpose of connecting businesses and artists (fusing culture and commerce) while raising awareness about the South End.

Resources and ideas for making it happen in your community

Revitalization 101: When revitalizing a run-down section of town, choose the most run-down property on the most traveled street to rehabilitate. If possible use troubled youth workforce training programs to work on redeveloping the property. Repeat the process.

Family-friendly places: Work collectively with the community to develop family-friendly activities for a successful downtown. Children usually trump their parents' concerns about fighting traffic or finding or paying for a parking spot.

Places to play: Clean up and fix up school grounds; build safe places for children to play.

Holiday parking: Ask private property owners to allow the public to use their parking lots for free on nights and weekends during the holiday season. It's a good way for businesses to be charitable at this time. Promote the free use to media and throughout the community.

Visual cues to improve Main Street: Dress up downtown with hanging flower baskets and uniform outdoor lighting. Work with property owners to improve their facades. Small things induce bigger projects over time.

5 I'm OK, You're OK: Social Well-being

How Are Civic-minded Organizations Supported for Fostering Social Well-being?

What does it take for a community to thrive? Not only in terms of economic or physical aspects but social aspects too? In most cases, these social aspects serve as the underlying foundation on which to build any sort of community and economic development policies or programs – it's the social economy sector that contributes vitally to the creation of a durable economy. This underlying foundation is what facilitates or leads to community development, and is generally referred to as social capital or capacity. Simply put, social capital or capacity is the extent to which members of a community can work together effectively to develop and sustain strong relationships; solve problems and make group decisions; and collaborate to effectively plan, set goals, and get things done.[1] The key to social capital lies in networks and connections in which people exchange benefits and assistance in nonmonetary ways. Civic engagement is crucial as part of this: people need to be aware and involved to effect positive change in their community.

How is social capital created or encouraged? The process of community development is social capital/capacity building which leads to the creation of social capital, which in turn leads to the outcome of community development.[2] Figure 5.1 shows this progression: note that solid lines show the primary flow of causality while feedback loops are represented by the dotted lines. As shown in the third box, progress in the outcome of community development (taking positive action resulting in improvements in the community) contributes to capacity building (the process of community development) and social capital.

It takes all types of organizations – public, private, and nonprofit – to help a community thrive. In the recent past, there has been a "blurring of lines" across these organizational types and sectors as societal needs and desires change, and new approaches are needed. Heerad Sabeti, of the Fourth Sector Network, describes the advent of a new generation of organizations at the intersection of public, private, and social sectors this way:

> (a) *The convergence of organizations toward a new landscape – a critical mass of organizations within the three sectors has been evolving, or converging, toward a fundamentally new organizational landscape that integrates social purposes with business methods;*
>
> (b) *The emergence of hybrid organizations – pioneering organizations have emerged with new models for addressing societal challenges that blend attributes and strategies from all sectors. They are creating hybrid organizations that transcend the usual sectoral boundaries and that resist easy classification within the three traditional sectors.*[3]

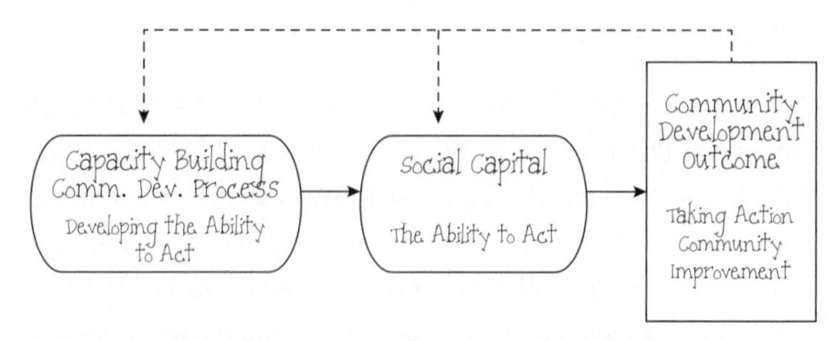

Figure 5.1 The process of community development
Source: R. Phillips and R. Pittman, Introduction to Community Development, London: Routledge, 2009, p. 7.

It's an exciting shift from traditional models to newer, more inclusive and socially conscious ways of thinking. It's seen in areas as wide ranging as social entrepreneurship, corporate social responsibility, green businesses, and mission-related investing. New forms of support are emerging too, from states enabling new corporate structures such as Vermont's B-Corp. legislation (Ben and Jerry's converted to a Benefit Corporation in 2012 – see Chapter 2 for more details on Benefit Corporations), to foundations and credit unions shifting funding to support more social enterprises. But perhaps most exciting (at least for those involved in public sector community and economic development) is that the public sector can engage in these activities. Going beyond traditional mindsets, a public agency can serve as a catalyst or change agent as a "fourth sector" force. Organizations embracing this approach are capable of connecting, facilitating, and supporting other government agencies, businesses, and the nonprofit sectors through strategic and business management while fostering social well-being. These "go to" organizations can get the job done with entrepreneurial, opportunistic approaches, and generally reflect the following principles:

- Relationships matter.
- Vision and mission are important but also must sustain the goals of building businesses and organizations or attaining funding.
- Money can be leveraged to build new businesses, programs, and services.
- Services can and should be connected.
- Strong business infrastructure serves to improve performance in all sectors.
- Continual organizational assessment and self-correction is vital to success (and dancing in the "end zone" is not).
- Data, research and informational technology can serve as the basis for service and program integration.[4]

Connecting to community = better health

Men's Health magazine recently examined the most socially networked cities in the U.S. Burlington came in at number 13, out of 100 cities and much larger competition such as San Francisco! See their website at www.menshealth.com/best-life/social-networking-cities, for more details.

Another article by *Men's Health* ranked the top ten cities for men's well-being in the U.S. Not surprisingly, those cities with a sense of friendliness and inclusiveness ranked higher and also more healthy for the men who live there. Their story references a study by Swiss researchers that points out those people with a supportive social network enjoy better cardiac health. It's those all important community connections we're talking about that have benefits beyond the obvious! We're proud to say that Burlington ranked fourth overall (and second in health, seventh in life expectancy). For more details, see the magazine's website at www.menshealth.com.

CEDO as change agent

It really started in the early 1980s when the City of Burlington asked, "What should CEDO do?" Residents were surveyed, and overwhelmingly, the answer was to guide development. From that day forward, it's been clearly driven by frequently asking the question of residents, "What do Burlington residents want?" And it's clear they want a fair, sustainable and vibrant economy while enjoying a high quality of life.

"We became an agent of change and a catalyst to help guide development vitality of all sectors," Bruce Seifer formerly of CEDO explains. "We're now what's called a fourth sector organization because we've always focused on how to facilitate and catalyze – leaders meeting on a whole range of topics, organizing groups to interact and become full members of the workforce, and so on." CEDO has found it is essential to give residents the opportunity to learn how to be effective community members – by involving them in public policy, neighborhood organizations, and civic engagement training for boards and City employees. There are so many different ways to engage people to facilitate and be a catalyst in the community, and training is vital for this. It goes beyond training though, it's also about empowering others to engage and pursue what they feel passionate about and what's important to the community. It's empowering to talk to others, seek out elected officials at all levels, and develop collaborative relationships. While social capital development is organic and evolves on its own in many aspects, having a catalyst organization such as CEDO has prompted a high level of social capacity development in Burlington.

In Burlington, strategic assessments are conducted to help find out what the community needs, whether it's local housing issues, technology, or the food sector – whatever it may be. Looking at trends and other data helps balance perspectives too. That's why *Jobs & People* is important to keep revisiting over the decades; it provides insights into trends and benchmarks, enabling an objective look at what's been accomplished and where the community needs to go in the future.

Building capacity via the social sector

CEDO and the community long ago realized the value of the social sector. It's part of the equity building process. These organizations are key, and early on, CEDO started providing funding and technical assistance to start them. Over time, many have found ways to become self-sustaining. Creating and supporting the mission of these organizations serves the constituencies, and builds social capital. It also builds assets in the community, and allows for dialogue to explore what's needed. This dialogue includes the opportunity to speak out, for constituency groups to interact with politicians, and to be part of a process for investing in each others' organizations. It becomes a mechanism for allowing "locals to invest in locals" and create relationships (and it works for the private sector too – allowing for capital to flow between community and businesses).

> *Our biggest strength is that we have a very engaged citizenry, where we are participants. While we all talk about democracy, there are lots of places where citizens feel like clients rather than participants in government. Combining this with values focused on relationships, action and healthy lifestyles and less on status and material goods gives us strength as a community.*
>
> (Betsy Ferries, former director, Mercy Connections)

Since 1984, CEDO has helped launch over 20 nonprofit organizations. Several of these "spun-off" from CEDO initiatives; others were aided with financial or technical support. Below is a list of these organizations. It's important to note that it could be argued that not all types of nonprofits necessarily generate social capital with broader community impacts. So it matters which types of nonprofits are targeted to develop and support – in other words, those that will help build social capital and capacity, to help foster community and economic development outcomes in the host area. Thinking strategically about which nonprofits can help with overall mission and goals, and supporting those, enables more capacity to be built, and in turn, assists in fostering a more durable economy.

Generating nonprofits

CEDO focuses on building nonprofit capacity and has helped launch the following organizations:

- 1984 Burlington Community Land Trust, now called Champlain Housing Trust
- 1984 Lake Champlain Housing Development Corporation (merged in Champlain Housing Trust)
- 1984 Burlington Youth Employment Program (now defunct)
- 1985 Step Up for Women, now called Vermont Works for Women
- 1985 South End Arts + Business Association

1986 Vermont Energy Investment Corporation
1986 Burlington Vermont Convention Bureau
1987 Women's Small Business Program
1988 Recycle North, now called ReSOURCE
1990 Champlain Valley Mutual/Cooperative Housing Federation, now part of Champlain Housing Trust
1991 Vermont Businesses for Social Responsibility
1994 Vermont Community Enterprise Fund
1995 Vermont Sustainable Jobs Fund
1995 School to Work Initiative, now called navigate
1995 ECHO Lake Aquarium and Science Center
1996 Good News Garage
1996 Community Health Center Dental Clinic
2001 Vermont Employee Ownership Center
2003 Vermont Software Developers Alliance, now called Vermont Technology Alliance
2004 Vermont 3.0 Creative/Tech Career Jam, now called Vermont Tech Jam
2007 Old North End Arts + Business Network
2008 Chief Information Officers Organization
2008 CarShare Vermont
2008 Vermont BioScience Alliance.

From the latest *Jobs & People* (2010), it's evident that the area benefits from efforts to build social capacity, and that factors combine here to foster their creation. Figure 5.2 presents data at the county level (the level of availability) and it's interesting to note that Chittenden County, of which Burlington is part of, has substantially more nonprofit organizations per capita than the nation as a whole, a gap which has been increasing steadily since 1995 with growth rates at times nearly double those of the rest of the country.[5] Table 5.1 shows the growth in registered nonprofits at local (Chittenden County and Vermont State) and national levels.

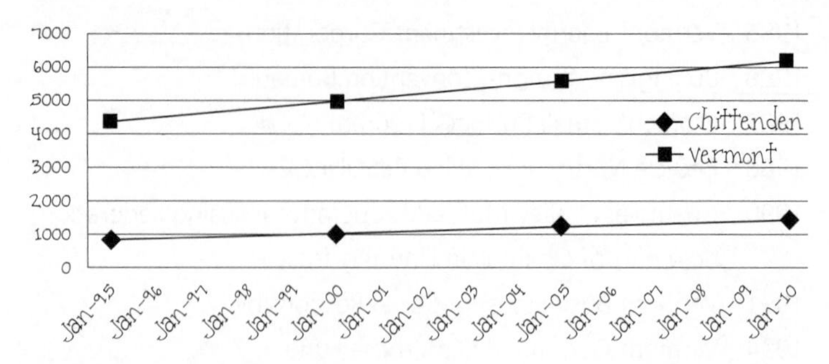

Figure 5.2 Registered nonprofits: Chittenden County, Vermont, 1995, 2000, 2005, 2010

Table 5.1 Registered nonprofits and their growth

	Registered nonprofits (n.)		
	Chittenden	*Vermont*	*U.S.*
August 1995	871	4,381	1,066,808
May 2000	1,038	4,897	1,211,937
November 2005	1,269	5,592	1,409,645
January 2010	1,437	6,221	1,581,825
Nonprofits per 10,000 people, January 2010	98.04	102.18	56.21
	% Growth in registered nonprofits		
	Chittenden	*Vermont*	*U.S.*
1995–2000	19.2%	11.8%	13.6%
2000–2005	22.3%	14.2%	16.3%
2005–2010	13.2%	11.2%	12.2%

Major nonprofit employers

Has the strategy worked? In some cases, yes! Several of the nonprofits launched with CEDO's help are now major employers for the area. Here's some info about this dimension of fostering social well-being and capacity. From *Jobs & People IV* (2010), it's noted that the majority of major Burlington nonprofit employers are involved in education, health care, or social services. It's clear that the organizations listed play an important employment role in the community, in addition to the services they provide. Fletcher Allen Health Care is by far the largest employer of this group, and the Community Health Center also employs a significant number of workers. While the University of Vermont does not fall under this category as it's considered a government employer as a public university, nonprofit Champlain College and Burlington College are also major employers. From Table 5.2 below, it's evident that a wide range of social service agencies, from nonprofits offering direct service, housing or legal assistance, to job training programs, are also included.[6]

Illustration: creating models for community well-being

Some of the nonprofits launched have done exceptionally well, having significant impact on a variety of fronts. One of these, the Champlain Housing Trust, started in 1984. With its creation, Burlington founded the first municipally funded community land trust in the U.S. It's now grown into one of the nation's largest, with over 2,500 voting members and an affordable housing and community portfolio with 650 units. The Trust's impact is felt beyond its own projects, providing key support for the area's comprehensive housing strategy. This land trust model of shared equity has widened opportunities for home ownership. They've also committed to creating permanent affordable units with approaches such as restrictions on resale. Another notable? More than two-thirds of participants have become homeowners, helped in part by their land trust equity gains realized during their time in the program.

Table 5.2 Major Burlington-based nonprofit employers, 2008

Nonprofit organization	*# Employees*
Fletcher Allen Health Care Inc.	6,877
Howard Center Inc.	1,017
Greater Burlington Young Men's Christian Association Inc.	690
Champlain College Inc.	505
Vermont Energy Investment Corp.	179
Vermont Catholic Charities Inc.	153
Spectrum Inc. Spectrum Youth and Family Services	130
Resource A Nonprofit Community Enterprise Inc.	121
Community Health Center of Burlington Inc.	117
Burlington College	109
Flynn Center for the Performing Arts Ltd.	94
Vermont Legal Aid Inc.	75
Champlain Housing Trust	73

Source: National Center for Charitable Statistics at the Urban Institute; 2008 IRS Form 990 images provided by The Foundation Center. Note: This list is not exhaustive and was generated looking at a list of all registered Burlington nonprofits sorted by Gross Receipts. Employment figures were calculated for the 50 largest organizations.

Figure 5.3 Champlain Housing Trust logo

For additional details, see their website at: www.champlainhousingtrust.org/

2010 – City of Burlington receives the 2nd annual Home Depot Foundation Award of "Excellence for Sustainable Community Development", small city category

Figure 5.4 Champlain Housing Trust Headquarters
Photo: Champlain Housing Trust.

Winner – Burlington, VT and Champlain Housing Trust

In response to increasing urban sprawl that posed a threat to economic and social vitality as well as the environmental health of the community, the City of Burlington, local residents, and businesses came together to identify and implement a strategic road map for the future called the Legacy Project Action Plan, which represents a holistic approach to creating a sustainable community.

Burlington Cohousing East Village is directly aligned with the goals of the Burlington Legacy Project Action Plan. Completed in 2007, Burlington Cohousing East Village is Vermont's only urban cohousing community. Cohousing offers private homes with shared common spaces, including community gardens, outdoor courtyards and laundry facilities. This 32-home development exemplifies best practices in housing, natural resources and land use and development.

(Source: www.homedepotfoundation.org/awards/scd/winners.html)

The role of the business community

Former mayor Peter Clavelle describes this role:

> *Businesses in our city have a direct interest in assuring that Burlington offers a skilled, high-quality workforce; creative financial incentives and other supports for new and existing businesses; and telecommunications and other infrastructure enhancements that help businesses work more economically, efficiently, and effectively. Business can take a leadership role in developing and supporting civic projects. The business community brings many resources to the Legacy Project table, from capital to technical expertise to providing training sites for city residents seeking to build job skills. Nonprofit social service, health care, environmental, and neighborhood organizations will play central roles in achieving the goals of the Legacy Project. Working closely with government, schools, and the business community, these organizations will be essential in implementing the actions of the Legacy Plan. Through services to their clients and members, financial resources, research and publication efforts, and community outreach efforts, nonprofit organizations can have a profound impact on all aspects of the Legacy Plan. These local organizations also have an important role in maximizing the strengths of the individuals they serve, helping to tap the cultural viewpoints, expertise, and energy of their constituents in the service of a sustainable Burlington.*[7]

What is it about Burlington and the area that attracts a socially focused and entrepreneurial crowd? Part of it is the history as an attractive area in the 1960s and 1970s for alternative culture. But it has to go beyond that, or it wouldn't have sustained itself over the ensuing decades. Burlington and Vermont are both noted for generating a certain climate, one that particularly attracts social entrepreneurs to create and grow fourth sector enterprises. Here's a few perspectives on this from local leaders:

I cannot tell you another area that has this culture. Why? Maybe because we had to do it, we like living here and we were desperate when we first got here. There is a liberal bias too of why we came here, so it's self-selection. And although none of us came here starting out in business, we survived and helped create its entrepreneurial climate.

(Alan Newman, a self-confessed unemployable insubordinate, helped found Gardener's Supply and Seventh Generation before launching Magic Hat Brewery)

The alignment between our company's mission and Burlington is good, our future is similar and we share values.

(Steve Conant, Conant Metal and Lighting, longtime entrepreneur and nonprofit leader)

There's a balance of sectors where everyone has a role and no one is dominant. It's where leadership can merge and elicit alignment with the continuum towards moving towards the "whole."

(Will Raap, Gardener's Supply Company, founder; founding member, Intervale Center)

Creating a legacy of good will

Burlington's Legacy Project and its plan have been mentioned before, and it's worth revisiting the 2030 vision for the community in the context of building social capital. Working with the civic sector is an instrumental focus of the Legacy Plan; comprehensive development requires that all sectors work together and across boundaries (both within the community and the larger region) to create and implement a community vision. Below are some of the activities and focus of these efforts as related to building on neighborhood strengths. These efforts help expand and encourage a resident-focused governance structure in Burlington.

It's a progressive city where citizens and visitors both want to be. And as long as there's a group of people here having lots to say with a sense of entitlement that they have a voice, it will be a citizen-led government structure.

(Robbie Harold, management consultant and former director of economic development and deputy commissioner of the Department of Development for the State of Vermont)

Legacy Project is honored as a "bright idea"!

Bright Ideas is an initiative that recognizes creative and promising government programs and partnerships. The initiative is offered through the Innovations in Government Program, a program of the Ash Center for Democratic Governance and Innovation at Harvard Kennedy School. For more information, see: http://innovationsaward.harvard.edu/BrightIdeas.cfm

---

Neighborhoods as building blocks of community[8]

As part of the Legacy Plan, the Burlington Neighborhood Project is a community-organizing initiative aimed at improving the quality of life in Burlington's lower-income neighborhoods. The following is an excerpt from the Plan explaining how Burlington is working with its neighborhoods to build resiliency and improve community well-being (see link to website listed in the Resources section for additional details).

*In the beginning, this effort was staffed by AmeriCorps*VISTAs[9] and supported by professional Community Development staff, with the project focused on establishing neighborhood associations, developing local leadership, and educating about crime prevention – through canvassing and various other outreach efforts. So far, the Burlington Neighborhood Project has helped to establish and support more than 50 street and block associations. Neighborhood groups are empowered through programs that fund block parties and neighborhood improvement* *projects, assist in the facilitation of meetings, and conduct free leadership trainings.*

Neighborhood Planning Assemblies (or NPAs) also bring together citizens from each of Burlington's seven Wards for monthly meetings. These forums offer people opportunities to learn about issues, make plans, and effect change through allocating City resources in their neighborhoods. Annual Neighborhood Improvement Nights at the NPAs provide additional opportunities for citizens to discuss and prioritize spending on municipal improvements.

Burlington also works to resolve disputes that sometimes brew among neighbors with different lifestyles and expectations. The Community Support Program provides a mediator to help residents resolve or manage neighbor-to-neighbor conflicts.

In the downtown neighborhood – and on the four-block-long pedestrian mall known as the Church Street Marketplace – another comprehensive, community-based solution is at work. A team of mental health outreach workers was established to care for those with mental illness and reduce disruptive behavior. This program came about through the efforts of the Downtown Action Group, which unites representatives of business and government. The group lobbied for and received state funding toward a full-time outreach worker delivering street-based services. Subsequently, a major grant was received to add personnel and expand their outreach responsibilities to include people with substance abuse or behavioral problems. Central to the successful initiation of this program was a business community willing to step forward and call for additional spending on human services.

Burlington's downtown neighborhood serves as the entertainment center for an entire region – and for students from the University of Vermont and other colleges. To address rowdiness, noise, excessive drinking, and other issues arising from the downtown bar scene, Burlington has joined cities across the country in establishing a Hospitality Resource Panel. This panel involves key stakeholders in developing strategies and policies to strengthen our hospitality industry and to maintain a comfortable quality of life for downtown's growing residential population.

Other initiatives to help strengthen Burlington neighborhoods have included Racism Study Circles, Substance Abuse Study Circles, a comprehensive handbook of neighborhood resources and regulations, welcome packets for new residents, and community newsletters and listserves.

Celebrating successes is important too. Burlington's Neighborhood Night of Success is an annual celebration of neighborhood activists, whose hard work and innovative ideas make Burlington a great place to live, work, and play. The event includes a free dinner, entertainment, kids' activities, and an award ceremony that honors community leaders and groups.

Another key component is Burlington's Community Justice Center. The center's goal is to provide direct participation of community members in delivering justice and holding those who commit nonviolent crimes directly accountable to their victims and the communities that have been harmed. Volunteer Victim Liaisons reach out to crime victims. Restorative Justice Panels help negotiate reparative agreements between victim, offender, and community. The center then monitors community-service hours for offenders.

Center programs focus on party noise, vandalism, low-level crime, and youth offenders. The Community Justice Center is also working to increase the accountability of people returning to Burlington from prison and to support their efforts to live productively in the community.[10]

The Legacy Project's Social Equity Investment Project focuses on identifying and supporting leadership development in the community. Its purpose? To facilitate sustainable and effective social change via capacity building – social equity focus groups, fostering a coordinating network, and financial development assistance. It's a social development resource created to help Vermont leadership and communities better address growing cultural diverse and socioeconomic issues which impact quality of life for all populations. The goal is to move Vermont beyond the challenging and transformational crossroads, recognizing commonality and a shared vision. It's gaining attention, even beyond Burlington. SEIP received the 2nd place 2009 City Cultural Diversity Award presented by the National Black Caucus of Local Elected Officials (NBC-LEO) for developing creative and effective programs designed to improve and promote cultural diversity through a collaborative process with City officials, community leaders and residents. The SEIP model will be placed in the National League of Cities database for national best practices.[11]

The total is greater than the sum of its parts

As seen above, it takes many partners across sectors to build community capacity and social capital, which in turn leads to desirable outcomes. It requires transcending the traditional or "business as usual" mindset so that organizations can serve as catalysts for positive change. Fostering and supporting these efforts while also functioning as a change agent in the public sector generates the proper environment for the social economy to thrive. Adding all these efforts together creates a community where it feels possible to tackle issues while building on strengths. It really does add up to something greater than the sum of its parts.

Our selection for the Closer Look case in this chapter centers on the Good News Garage, a nonprofit organization that's received national attention for its innovative approaches. It's interesting to note the strong connections inherent in this case: CEDO provided technical assistance to start it up, and it occupies a former brownfield site, that subsequently received an EPA award as a showcase project.

A closer look: the Good News Garage

Figure 5.5 Good News Garage logo

In 1996, CEDO provided assistance to help launch the Good News Garage, now a nationally acclaimed model for its success at meeting two critical community needs: skilled job training and affordable, reliable transportation. Using $10,000 of CDBG funding from the City and a $35,000 grant from Lutheran Social Services, the organization started with the concept of donating cars to needy single mothers (at the time, 80 percent of those in poverty were single mothers). Transportation equity was the primary focus behind its creation, and helping those who would benefit greatly from having their own car.

Hal Colston, the founder, explains the success of the program:

> *It really comes down to relationships – how do we get to understand and get to know someone who is different from me? Asking people what they need, not what you think they need. Focus on what you can control and let go of what you cannot. At that point, stakeholders and others will appear!*

How did being in Burlington help? It's a human-scale city, and this made a difference. "If you show up and want to be involved, you're in," Hal says. "It's also a place where there is a preponderance of nonprofits and people who value social capital, and this is instrumental to forging relationships that can benefit all."

The program, which has now spread to other New England states, solicits donated vehicles, trains people to recondition them, and provides the vehicles to people in need at the cost of the repairs.[12] Their mission is to "promote economic opportunity through our training program while providing dependable private transportation that in turn helps people succeed in the job market." Founded in 1996 as part of Burlington's Old North End Enterprise Community, the program has awarded approximately 4,000 reliable vehicles, to individuals and families in need. Working with trainees referred by the Vermont Department of Labor and Social Welfare, the Good News Garage has a 100 percent job placement record to date. Half of the program's trainees have been women. To be eligible to receive a Good News Garage vehicle, individuals must possess a valid driver's license and earn less than 150 percent of the federal poverty level. "When we have to prioritize, we focus on the person who is already employed, whose car has died, and who needs a reliable vehicle to get to work," Colston points out. "We also look at whether the person is about to start a job and needs transportation or needs a car in order to take part in a job training program." Seventy-five percent of the program's vehicle recipients find work and move off the welfare roles. "Having reliable, affordable transportation makes all the difference," Colston notes, "It blows away the whole notion that people don't want to work."

Good News Garage has received a multitude of media attention, including an appearance on *Oprah*. All the attention began with a piece about moving people from welfare to work on Vermont Public Radio. This was picked up by National Public Radio's *All Things Considered*, then others followed suit, from *NBC Nightly News with Tom Brokaw*, and *CBS Evening News with Dan Rather*, to the *Smithsonian* magazine.

It's more than media, however. Its measurable impacts are impressive, and the program is making a difference. Here's an excerpt from the impacts reported on the Good News Garage website:[13]

The problem

Access to transportation is a major barrier for low-income women and others trying to gain economic security.

Our qualified applicants have a significantly difficult commute, are faced with non-standard shifts, require multiple childcare drops, or have jobs that require additional work-related travel or a reverse commute.

Lack of a vehicle also limits access to affordable housing, healthcare, education, childcare, shopping and religious services.

Lack of a vehicle consumes significant time for parents and their children – time that could be spent working, studying, participating in after-school activities, or investing in family time. Parents are not well-positioned to care and provide for children without a vehicle in today's American society.

More than 80% of clients served are single parents who are struggling to achieve financial independence and a better quality of life for themselves and their children.

Results

Based on a 2006 Impact Study conducted by the University of Vermont and the U.S. Department of Housing and Urban Development (HUD), the following statistics were discovered:

- *A total of 61% reported a decrease in their reliance on public assistance (Temporary Assistance for Needy Families, or TANF) due to the vehicle.*
- *The majority of the people who had decreased their reliance had done so completely, that is, 49% of the population reported a complete decrease in their reliance on TANF due to the vehicle.*

- *37% of the population reported a decrease in food stamps due to the car.*
- *60% of the population attributed the obtainment of employment to the car.*
- *83% of the population attributed the ability to keep a job to the car.*
- *58% of the population reported an increase in some sort of community participation due to the vehicle;*
- *48% of the population attributed an increase in education to the car.*
- *60% of the population attributed an increase in training to the car.*
- *90% of the population reported an improvement in hope for the future of themselves and their family members within Vermont due to the car.*
- *87% attributed an increase in self-confidence due to the car.*

Resources and ideas for making it happen in your community

City of Burlington, *Jobs & People IV,* www.cedoburlington.org/business/J&P_IV/j&p_iv.htm

Good News Garage, www.goodnewsgarage.org/

Legacy Plan Project, http://burlingtonlegacyproject.org/

Community transportation: Meet with transportation officials and share resources to develop bus services to new areas where employers are locating. Create car share or car donor programs to help people get off public assistance and get jobs. Fix up and maintain walking routes to schools; and/or create councils to solve transportation and parking problems.

Create a council of service providers: Create a council of small business service providers to share information on your programs and services. Bring in outside service providers to learn about their services; include small business development center, a service corps of retired executives, community action, small business administration, community colleges, universities, bank loan officers, local government. Meet at least four times a year.

Youth activism: Get youth appointed to community boards as voting members or as youth advisors to build future leaders. Commit to carry out one of their ideas to inspire the youth in your community.

Connect with community foundations: Meet with staff from the local community foundation to find projects to work on collaboratively. If such a community foundation doesn't exist, work to establish one.

Capital ideas: Train volunteers and staff on loan underwriting. Use and leverage local revolving loan funds. Many communities have underutilized loan funds that can leverage other resources to grow your economy.

Maximize resources: Organize a meeting for all the service providers in a particular issue area like youth advocacy. Identify problems or issues, opportunities and next steps. Provide a sign-up sheet to follow up on next steps.

Facilitative leadership: Host trainings in facilitative leadership to train government and nonprofit officials in how to maximize their time and effort.

6 Glowing and Growing: Energy and Environment

How Is Ecological Balance Maintained, and Efficient Energy Sources Provided?

Community energy issues and policies impact development outcomes, and this is certainly an influential factor in the environmentally sustainable economy subset. This is observed in regions and cities all over the globe. Reliable, efficient and affordable energy is central to long-term sustainability. Energy generation has direct impacts on environmental quality, depending on the type and intensity. Even sources such as solar or hydro convey environmental consequences. Targeting less impactful energy sources fosters sustainability and protects environmental quality. It's not just about generation or sources of energy, it is about usage and conservation too. Most economic models operate in a linear fashion, with various assumptions about unlimited supply, inputs, increasing consumption, and continuous growth. However, it's increasingly clear that we have to consider how finite our sources and inputs are, especially in regards to energy. The gradual realization that our natural system does indeed have limits needs to translate directly into effective energy policies.

Burlington is tackling this challenge using conservation measures as well as considering applications such as a district heating system. This system would capture the excess heat from the city's 50-megawatt biomass (wood chips) electric generation plant, sending hot water out to heat over 5,000,000 square feet of buildings and homes. Greenhouse gas emissions in Burlington would be reduced by 10,000 tons per year. The cost to build a district energy system is high, and is influenced by public policy that currently is not conducive to bonding for the infrastructure, or federal policy that doesn't yet support this type of approach. It's been an ongoing issue for years, and remains a challenge for long-term energy provision and efficiency.

In the meantime, conservation rules. Since 1990, there's been a commitment to conservation, and it's paid off as electric usage 20 years later is only 1 percent more, despite the increased use and availability of products such as cellphones, laptops, etc., that use electricity. Instead of expanding generation capacity, an $11.3 million bond was issued to support conservation by providing incentives and support for energy-efficiency measures; it included conforming new construction to Leadership in Energy and Environmental Design (LEED) green building rating system. LEED ratings, developed by the U.S. Green Building Council, require specific standards for construction and retrofitting of buildings. This chapter provides information on several programs and initiatives to foster energy conservation and efficiency. Our Closer Look selection is the case of the Vermont Sustainable Jobs Fund.

Property Assessed Clean Energy program (PACE)

Other initiatives include requiring buyers of multi-family apartment buildings to bring those properties up to energy codes. They have 12 months after time of sale to do so. This has helped increase efficiencies, especially for older properties in need of retrofitting. Act 45, passed in 2009 by the Vermont State Legislature, included provisions that permit communities to establish Property Assessed Clean Energy assessment districts for the purpose of enabling property owners to invest in energy efficiency and/or renewable energy projects in existing homes. This program will be run by Burlington Electric Department. The POWER program (Property Owners Win with Efficiency and Renewables) will provide a special tax assessment district created by the City, in this case all properties located within the City limits. The City would be responsible for establishing a fund for the district with participating property owners accessing financing for eligible energy efficiency and renewable energy projects. Funds (these are not loans) made to approved property owners would be paid back as an assessment on their property tax bill. Note that this repayment transfers to a new property owner at the time of sale if buyer and seller agree.

The cost of the PACE program will be borne solely by the participating property owners, as stated by law, and not as an added cost to City services. Repayments are calculated to recover the costs of the program over a period of time that is no longer than the useful life of the installed efficiency measures or renewable energy project, weighted by cost.

There are numerous types of efficiency and renewable energy measures qualifying for this program. Renewable energy is defined by Vermont state law as "energy produced using a technology that relies on a resource that is being consumed at a harvest rate at or below its natural regeneration rate," specifically including "flammable gases produced by the decay of sewage treatment plant wastes or landfill wastes and anaerobic digestion of agricultural products, byproducts, or wastes," but excluding "solid waste, other than agricultural or silvicultural waste," any "form of nuclear fuel" and hydroelectric energy from a plant over 200 megawatts. These can include solar heating systems, biomass energy systems, wind and micro-hydro applications. Eligible efficiency measures are defined in a list published annually by Efficiency Vermont and the Burlington Electric Department. If a City-wide district energy system is built, the costs carried at the commercial and residential property level could be financed through the Clean Energy Assessment District.

Prior to a property owner receiving PACE financing, an analysis of the proposed project has to be performed. As a Vermont-designated Energy Efficiency Utility, the Burlington Electric Department performs these analyses, gauging project costs and energy savings as well as estimating carbon impacts of the proposed energy improvements, including an annual cash-flow analysis. The Department explains,

> *The PACE-financed portions of residential energy improvements cannot exceed $30,000 or 15 percent of the assessed value of the property. The loan-to-value ratio (any outstanding mortgages plus the amount of PACE financing) cannot exceed 90 percent of the assessed property value. For commercial properties, PACE financing cannot exceed 15 percent of the assessed property value and the loan-to-value ratio (any outstanding mortgages plus the amount of PACE financing) cannot exceed 90 percent of the assessed property value.*[1]

Unfortunately, just as the program was about to rolled out to the public, the national banking/mortgage industry successfully lobbied to stop PACE programs around the country. Because the assessment functioned as the equivalent of a tax lien, in the event of a default, repayment of the PACE assessment would technically be collectable by the municipality before the mortgage lien could be exercised. This put Burlington's PACE program on hold for over a year until the State of Vermont's Department of Financial Regulation issued new standards for both program regulations and underwriting. At the time of publication of this book, Burlington's PACE program was still on hold but Burlington Electric Department is optimistic that a residential-only version will be rolled out in the near future as a trial.

Burlington Electric Department announced in October 2012 that they are starting an energy-efficiency pilot program funded in part, with a grant from the Economic Development Administration that will include elements of the PACE program for small businesses. Burlington Electric will offer upfront funds to customers to make energy improvements, with the resulting savings used to pay back the loan over time in installments on the customer's energy bill.

This new on-bill finance program will help 1,500 businesses each access $6,000 to make efficiency improvements, providing a payback in four years.

October 10, 2012: BED to start on-bill financing for energy-efficiency projects

U.S. Senator Bernie Sanders (I-Vt.) recently announced a new $1 million grant to support energy efficiency in Burlington. "This program may well become a national model for how we overcome hurdles to energy-efficiency investments. It will help us reduce energy costs to make our businesses more competitive, support jobs as we retrofit our buildings and reduce greenhouse gas emissions," said Sanders, a member of both the Senate energy and environment committees. Authorized by a Sanders provision in a 2011 energy bill, the U.S. Economic Development Administration grant will help Burlington Electric Department provide innovative on-bill financing for energy-efficiency projects, initially focused on small businesses. What Sanders called the "common-sense concept" behind on-bill financing is that utilities offer upfront funds to customers to make energy improvements, with the resulting savings used to pay back the loan over time in installments on the customer's energy bill. As chairman of the Senate's Green Jobs and New Economy Subcommittee, Sanders last year held a hearing to examine on-bill financing strategies. "Today's grant announcement is great news and builds on Burlington's continuing commitment to improve energy efficiency and keep energy costs affordable for businesses in our community," Burlington mayor Miro Weinberger said. "This award will further our efforts to make Burlington a leader on sustainability."

Burlington is once again showing its energy leadership by launching on-bill financing for its commercial customers. "The state's Comprehensive Energy Plan recognizes that on-bill financing is a valuable tool to help customers make efficiency investments, saving energy and money," said Elizabeth Miller, commissioner of the Vermont Department of Public Service. "This past year, the department worked with BED to include its on-bill financing program in its program performance review, and so we are very pleased that this funding will allow BED to accelerate its efforts, bringing greater benefits to its customers." Tom Buckley, BED's manager of Customer and Energy Services, added, "It's great that we'll be able to use this grant to address the small business customers who have traditionally had a hard time taking advantage of our efficiency program offerings."

After factoring in BED's energy-efficiency incentives, the average cost of an energy-efficiency retrofit for a participating business is estimated to be about $6,000. The investment will yield an average estimated savings of approximately $1,500 per year, providing a payback in four years. BED hopes to work with 1,500 businesses over the next several years. Today, only about 40 percent of the businesses in Burlington that undertake an energy audit actually move forward with energy-efficiency investments. This new on-bill finance program will help more businesses access funds to make efficiency improvements.

Vermont Energy Investment Corporation, in partnership with Vermont Housing and Conservation Board and Champlain Housing Trust, has received a $350,000 grant from the Doris Duke Charitable Foundation. The two-year grant will be used to demonstrate how deep energy-efficiency retrofits in single- and multi-family residences can make housing permanently and comprehensively affordable by reducing energy usage and costs (www.champlainhousingtrust.org/news).

Vermont Energy Investment Corporation (VEIC)

The Vermont Energy Investment Corporation (VEIC) is a mission-driven nonprofit organization founded in 1986 and dedicated to reducing the economic, social, and environmental costs of energy consumption through cost-effective energy efficiency and renewable energy technologies. VEIC develops energy policy and advocates for its passage in local, regional, national, and international forums. Beth Sachs, VEIC's founder, explains its genesis:

> *There were five different organizations in Burlington all interested in energy – but none needed a dedicated staff. In 1985 we started working with CEDO, Burlington's municipal electric utility, the local housing authority, the local community action agency and the state finance agency to create a nonprofit energy group that could perform a variety of high-quality energy services for all of these organizations. Then in 1999, the Vermont Legislature empowered the state's regulatory agency to create Efficiency Vermont, the nation's first "energy efficiency utility," which VEIC won the opportunity to operate through a competitive bidding process. This unique approach to delivering energy efficiency is inspiring to other entities in the U.S. and Canada, and VEIC has a very active consulting group that works with other cities, utilities, states, and countries to identify efficiency*

approaches and plans. The long-term results of VEIC's efforts have been impressive. Starting as a two-person effort, VEIC has now grown to a $60-million organization with nearly 275 employees.

Efficiency Vermont

Efficiency Vermont's accomplishments

Electric energy efficiency savings

As a result of energy-efficiency investments made from 2009 to 2011, Efficiency Vermont has helped ratepayers reduce their annual electricity usage by approximately 297 million kWh. Cumulatively, efficiency measures installed since the creation of Efficiency Vermont in the year 2000 provided 11.5 percent of Vermont's electric energy requirements in 2011.

Heating and process fuel ("thermal") efficiency savings

Beginning in 2009, Efficiency Vermont began combining its Forward Capacity Market revenues with Vermont's Regional Greenhouse Gas Initiative proceeds to offer a variety of services and incentives that result in improved building performance and heating system efficiency, thereby creating jobs, saving Vermonters money and reducing Vermont's greenhouse gas emissions. During those three years, over 3,000 customers have been served with total energy savings of approximately 87,092 MMBTUs.

Financial savings

The energy efficiency investments made by Efficiency Vermont lead to financial savings for all Vermonters. Electric rates are a reflection of utility costs. When Vermonters save energy, utilities generally need to buy less energy. As a result, utility costs associated with buying energy are less than they otherwise would be, and therefore the rates paid by all consumers are less than they otherwise would be. Efficiency Vermont has consistently spent less to purchase each kilowatt-hour (kWh) of energy efficiency than it would have cost Vermont's utilities to purchase the same kWh in the New England wholesale power market and deliver it to customers.

For example, in 2011, the levelized cost of Efficiency Vermont's total expenditures was approximately 4.8 cents per kWh. Taking into account participating customers' additional costs and savings, the levelized net resource cost of saved electric energy was 1.6 cents per kWh. To supply the same energy and capacity over the average ten-year life of efficiency measures installed in 2011, Vermont electric utilities would have to spend approximately 11.2 cents per kWh, based on current values of avoided costs.

The customers who have worked with Efficiency Vermont experience additional significant savings as a result of their reduced electricity consumption.

In total, Vermonters will save more than $309 million over the life of the efficient products and practices Efficiency Vermont put in place since 2011.

Environmental benefits

These same energy-efficiency investments will also benefit Vermont's environment by reducing the demand for the production of energy that emits greenhouse gases. Energy-efficiency measures installed from 2009 to 2011 will, over their lifetime, eliminate hundreds of tons of air pollutants and over 2 million tons of carbon dioxide, the equivalent of taking 355,759[2] cars off the road for a year.

Efficiency Vermont highlights

See Efficiency Vermont's *2011 Highlights* and *Annual Report 2011* for more information about its accomplishments.

Reviews of Efficiency Vermont's costs and savings

Efficiency Vermont's costs and savings are subject to rigorous review and verification through an independent monthly review of all invoices, an annual independent financial audit, an annual savings verification process conducted by the Vermont Department of Public Service, and by a legislatively-mandated independent audit of savings and cost-effectiveness that occurs every three years. See the Board's webpage on oversight activities regarding the EEU Program for additional information about these activities.

Source: www.efficiencyvermont.com/docs/about_efficiency_vermont/annual_reports/2011-Annual-Report.pdf

Creating the legacy

Burlington has a long commitment to and history of environmental ethics. Energy and resource conservation is a big part of this and connects directly to sustainable community and economic development processes and outcomes. This work was honored in 2010, when Burlington received an award for excellence in sustainable community development from the Home Depot Foundation. (See the summary case at the end of this chapter.) But sustainability is about more than just the environment. Burlington embraces what's referred to as the four E's – equity, economy, education, and the environment. This approach is reflected in the

Legacy Action Plan, adopted by City Council in 2000 and serving as Burlington's 2030 vision.

Jennifer Green, Coordinator of the Burlington Legacy Project, explains the overarching approach:

> *Burlington continuously seeks opportunities to better manage and protect our environmental resources, while working on quality education for all. We also believe that a strong local economy is paramount and that our ethnic and cultural diversity is celebrated as an asset and resource. In all the work we do, we believe in increasing participation in community decision-making.*

Progress towards goals in all four areas is tracked via the report card, using summary indicators to quickly see status and conditions.

Another important sustainability initiative is the Climate Action Plan, first adopted in 2000, and now updated. This plan serves as a blueprint for guiding environmental action and policy with an eye towards reducing greenhouse gas emissions. The updated plan includes the voice of dozens of community-generated recommendations on how to reach target goals. Here's a sampling of several major areas of focus that have emerged:

1. *Waste reduction in City Hall*: Waste Reduction including waste reduction in City Hall through increased recycling and composting. Also, an Environmentally Preferred Purchasing Policy was created by the mayor's Green Team composed of representatives of each City department. Former executive secretary Jessie Frank took the lead in implementing the policy by coordinating office supply purchasing. As a result of this work, Jessie received the Governor's Award for Environmental Excellence and Pollution Prevention.
2. *Energy-efficiency and conservation block grants*: The City allocated some of its block grant funding to projects having

energy impacts; recent projects include district heating feasibility analysis and energy-efficiency improvements in Fletcher Free Library.

3 *Burlington's No-Idling campaign*: This entails a wide-scale outreach effort to encourage the public to comply with the No-Idling ordinance. Why idling? Because vehicle engine idling contributes to air quality decline and it's an easy way to help reduce emissions. This project has involved an extensive outreach strategy, including the placement of no-idling signs around the community. The Department of Public Works (DPW) Commission approved changes to the original ordinance, including limiting idling from 5 to 3 minutes and eliminating the winter season exemption. Legacy is currently working with a school-coordinator on outreach and education about idling in the school community.
4 *Solar on schools*: The Burlington School District, Burlington Electric Department (BED) and CEDO are collaborating with Encore Redevelopment in a public/private partnership to develop between 1.5 and 2 megawatts of PV power on the rooftops of 5–6 school buildings. As a private entity, Encore can access a number of federal subsidies that would not otherwise be available to the City or School District. Key to this effort is a Power Purchase Agreement being developed between BED and Encore that will provide compensation to the project equal to the costs avoided by BED. The project will "green" the schools, expand BED's renewable energy portfolio, and provide lease payment income to the School District in an amount around $44,000/year.

The Climate Action Plan helps foster interest in and response to environmental issues. It requires collaborative efforts and needs much attention. We know the reasons behind it – global climate changes eliciting changes that are difficult to ignore – and the need to figure out how to respond at the local level. This plan presents ideas and proposed actions to help address this issue.

Burlington has a long history of climate change planning, beginning in 1996 when Burlington became one of the first cities to join the "Cities for Climate Protection" campaign. This led to a 1998 City Council resolution to reduce emissions to 10 percent below 1990 levels and the creation the first climate action plan. Burlington conducted its first greenhouse gas inventory in 2008, which was accompanied by a community process that researched and brainstormed action items to reduce the City's climate footprint. Over 200 recommendations resulted, based on seven key themes:

- Energy efficiency in buildings
- Renewable energy resources
- City government transportation
- Community transportation
- Waste reduction and recycling
- Local farms, gardens, and food production
- Urban forestry and carbon offsets
- Policy and education.

These strategies were then sorted and filtered, analyzed, and prioritized, forming the basis of the climate action plan update.

> The Renewable Energy Atlas of Vermont allows users to identify, analyze, and visualize promising locations for renewable energy investments and is a valuable tool for transitioning toward a more sustainable state. The Atlas is a user-friendly GIS-based website where Vermonters can click on their town (or several towns or county/counties) and select from a range of renewable energy options. A map and analysis appears on the screen and both can be saved and printed.
>
> (www.vsjf.org/projects)

Support biomass-fueled district energy, and other renewable technologies

The climate protection benefits of district heating – particularly of a system fueled by sustainably harvested biomass – are greater than any other single measure considered by the task force. Potential emissions reductions are estimated to range from 10,000 to 15,000 tons of CO_2 annually. Discussions of district heating have begun to expand beyond the current key players, to encompass the larger community that will benefit from its establishment. By continuing to inform and engage the community in future discussions, the city can help build consensus on the complex issues that surround a project of this size. Another component is the City's support for the development of additional renewable or other climate-friendly energy projects. These could, for example, include a new or continuing commitment of resources and support for select demonstration installations (e.g. waterfront wind energy), and ongoing research into promising technologies (such as biomass gasification, biological wastewater treatment).

The McNeil Generating Station is a 50-megawatt biomass (wood-fired) electric generation facility and the potential test site for using excess heat for a district heating system for the city. The issue of capturing excess heat from the plant has been an ongoing one, and still awaiting resolution.

Implement transportation demand management projects, and support climate-friendly transportation policy

Transportation – specifically motor vehicle use – produced roughly 30 percent of Burlington's total CO_2 emissions in the past. To help reduce these, the city supports the implementation of strategies for transportation demand management (TDM). It also calls for political support for important, climate-friendly transportation policies at the regional, state and federal levels. Select TDM measures, such as further expansion of park-and-ride lots, can provide between 9,000 and 16,000 tons of annual emissions reductions. Additional emissions reductions are obtainable through policies such as increased minimum vehicle efficiency standards, and the shifting of hidden transportation costs to motor fuel taxes.

Brownfields redevelopment

Brownfields essentially are properties with real or perceived contamination at levels considered hazardous to human health and/or the environment – most brownfields are identified during real estate transactions. Burlington's Brownfields Program has been granted over $500,000 by the EPA, cleaned up 30 sites, and has been named a Showcase Community Finalist twice. The assessment, cleanup and redevelopment of brownfields are high priorities for the City of Burlington. Why? Nick Warner, special projects manager overseeing the Brownfields Program with the City, explains: "Because brownfields site conversions can improve ecological conditions, reduce risk to human health and the environment, increase the tax base, create new jobs, create new green space, and curb sprawl." Given Burlington's long history, there are plenty of opportunities to pursue brownfields redevelopment. Several of these center on the waterfront, for example, the Moran building that has long been the object of controversy – occupying a favorable waterfront site – is now the focus of a major redevelopment effort.[3]

Partnerships are instrumental for implementing brownfields redevelopment projects. Case in point: the Vermont Transit Bus Barn was renovated with funding and logistical support from the Brownfields Program, in collaboration with the Burlington Community Land Trust and Housing Vermont. One of the buildings is now home to the nonprofit Good News Garage (our Closer Look case selection in Chapter 5). Overall, this $5 million project created 25 units of new housing, 15,000 square feet of commercial

Figure 6.1 Oil bollards, in Lake Champlain, slated for removal. At one time, petroleum products were delivered to Burlington by barge traveling through the Champlain Canal. Waterfront uses in Burlington have included industrial, fishing and other activities through the decades.
Photo: Nick Warner.

space, a gateway public park, and has been named a U.S. Environmental Protection Agency (EPA) "success story." The following is a project summary provided by the EPA.

Brownfields transformation: the Vermont Transit Bus Barn[4]

> *The property at 343 North Winooski Avenue in Burlington, Vermont (also known as the "Vermont Transit Bus Barns" property) was successfully redeveloped as part of Burlington's work with the EPA Brownfields Assessment Demonstration Pilot Program. In 1996 Burlington's Community and Economic Development Office received $200,000 in funding from EPA's Brownfields Program. In both 1998 and 2000, Burlington was recognized as a finalist in the National Brownfields Showcase Communities competition and has received a total of $300,000 in supplemental funding. Work on the N. Winooski Avenue site*

commenced in fall of 1998. Completion of the entire project is anticipated for September 2001. It is a highly visible, "gateway property" to Burlington's Old North End. This Enterprise Community is the most densely populated and lowest-income area of the city and the state of Vermont. The property is 2.6 acres and had 40,000 square feet of enclosed space. The site has been continuously in use as a transportation center since 1885. In 1998, Burlington Rapid Transit Inc. owned the site and leased it to Vermont Transit, which is now owned by Greyhound. The facility engaged in routine activities such as oil changing, coolant flushing, washing of buses and parts, and tire repair.

In 1999 Vermont Transit decided to relocate its operations and Burlington Rapid Transit Inc. sold the property to Burlington Community Land Trust (BCLT), a 501(c)(3) non-profit development organization. BCLT controls the property in limited partnership with Housing Vermont as Bus Barns Housing Limited Partnership (BBHLP). BCLT also partnered with the city of Burlington as co-developer of the property. The Phase I environmental site assessment (ESA) was completed in fall of 1999 by ATC Associates and a full Phase II ESA was recommended. The assessments and development of the corrective action plan cost approximately $80,000. Using EPA Brownfields funds, Burlington paid $60,000 and BBHLP paid the balance. All remediation work has been and will continue to be paid for by BBHLP. Contaminants included asbestos, petroleum hydrocarbons, and a variety of compounds commonly associated with fleet maintenance and repair. Concrete slabs and the contaminated soil were removed from the site by SD Ireland. Ongoing work includes asbestos abatement, interior wall cleaning, and additional soil remediation.

The site had three principle structures; two historic brick barns and one steel-frame garage. The historic brick barns were renovated while the steel-frame garage was demolished to provide green space. One barn was converted into affordable apartments. Combined with a newly constructed building, 25 rental units of permanently affordable housing will be created. Burlington is currently experiencing an extreme housing shortage. In 2000, the rental vacancy rate was approximately 0.3 percent. Over the last ten years, an average of only 32 new rental units per year have entered the market. The new housing at the Bus Barns site was fully funded by sources including Vermont Housing and Conservation Board, the city of Burlington, and Low-Income Housing and Historic Tax Credits. The second brick barn was converted for commercial space. One occupant is the Good News Garage, which provides donated cars to low-income families for the cost of repairs only. The garage also includes a training program for low-income people interested in becoming mechanics. The first funding for this project started with a grant written by CEDO for several associated projects involving the Department of Public Works. Through a grant from the Metropolitan Planning Organization aimed at improving public safety, streets and sidewalks were redesigned in the area, changing traffic flow at a dangerous intersection, and enacting traffic calming measures.

(www.epa.gov/region1/brownfields/success/burlington_vt.html)

Housing rehabilitation, energy efficiency and brownfields

Housing is a major focus of the Brownfields Program in Burlington. There is a large stock of older residential buildings throughout the city, many in need of remediation, retrofitting and bringing up to code. Brian Pine , assistant director for housing with CEDO, explains,

We're striving to have a balanced housing market, with about a 50–50 balance of rental and owner-occupied units

(currently at 42 percent–58 percent) – key to this has been the "recycling" of properties with heavy participation from the nonprofit sector. The Champlain Housing Trust serves a vital role, helping bring properties up to standards and recycle scarce housing resources.

Similar to other cities, this balance of rental to owner-occupied housing units is a concern, and requires creative approaches. It's also about balancing costs versus income, and Burlington strives to foster affordability of housing in relation to wages. "Our housing policies center on four areas: (1) preservation of existing units; (2) production of new housing where appropriate; (3) protection of vulnerable residents' housing; and (4) promotion of home ownership," says Brian.

It's a balancing act, and brownfields redevelopment offers a chance to impact several of these areas by bringing units to an improved standard. In the Old North End of town, where income is lower and diversity is highest, we are trying hard to recycle older buildings into more productive uses, providing higher quality housing and commercial opportunities for residents.

Affordability? We control through social policy versus rent control, and by making more units available. Our housing policies have made a difference, we have a larger amount of affordable housing than most cities – 25 percent of our housing stock.

(Michael Monte, chief operating officer of the Champlain Housing Trust)

In 1984, Burlington Community Land Trust and Lake Champlain Housing Development Corporation were founded by the City of Burlington to provide affordable, safe, and decent housing to families and individuals with low to moderate incomes. As geographic territory, services, and funding sources increasingly overlapped, the two organizations decided to combine their assets and resources in 2006 into Champlain Housing Trust.

The Champlain Housing Trust is the recipient of a 2008 World Habitat Award at UN-HABITAT's global celebration of World Habitat Day. The award, presented by the Building and Social Housing Foundation in conjunction with the UN agency, recognizes the Champlain Housing Trust's permanently affordable housing programs as innovative, sustainable and transferable (www.champlainhousingtrust.org/news).

Burlington's Old North End housing brownfields revitalization projects use a partnership approach, combining federal, state, city, nonprofit and traditional funding sources. It requires close collaboration across sectors and organizations. This includes working with the Vermont Housing and Conservation Board that played an instrumental role in securing funding. Featured in the U.S. Environmental Protection Agency's Land and Community Revitalization, the Old North End project is detailed in the following box, providing a summary of how the recent 13 multi-site project was structured.[5]

Land and community revitalization

Old North End residential properties, Burlington, Vermont

Property details

Property address:	1322 Saint Paul St.; 299 N. Winooski Ave.; 27/31 Peru St.; 22/36 Johnson St.; 52–56 N. Champlain St.; 57–63 N. Champlain St.; 73–75 Sherman St.; 104 Intervale Ave.; 221 Pine St.; 194 Hickok St.; 88 Sherman St.; 36 Convent Square; 112–114 Archibald St.
Property size:	*2.18 acres*
Former use:	*Multi- and single-family residential homes constructed mostly in the late 1800s and early 1900s*
Contaminants found:	*Lead and asbestos in some units*
Current use:	*Affordable residences*
Current owner:	*Champlain Housing Trust*
Project partners:	*Champlain Housing Trust, Housing Vermont, City of Burlington.*

Funding details

EPA Brownfields Assessment Grant (2009):	*$9,126 used of a $200,000 Hazardous Substances Assessment grant*
Vermont Housing and Conservation Board:	*$1,157,400*
Burlington HOME program:	*$325,000*
NeighborWorks:	*$500,000*
Vermont Weatherization program:	*$225,000*
Burlington Lead program:	*$140,000*
Tax credit equity:	*$2,175,919*
Bank loan:	*$700,000.*

Project highlights

- Concerns about endemic contamination in the area were alleviated by Phase I Environmental Site Assessments, which revealed cleanup was not required except for lead and asbestos mitigation during renovations;
- Energy efficiencies will be up to the standard of new housing, while maintaining the historic structures;
- Renovations will provide affordable housing, improvement of the neighborhood and quality of life;
- Historically appropriate renovations to maintain and enhance the existing character of the neighborhood.

Drivers for redevelopment: The Champlain Housing Trust (CHT) has owned and managed these 13 properties in the Old North End (ONE) neighborhood for over 15 years. Due to the age of the structures (most are over 100 years old) and small size they have not been operated very efficiently or in a manner that would allow their ongoing capital needs to be adequately funded. The CHT assembled these buildings and is in the process of selling the properties to a new tax credit partnership, City Neighborhoods, in order to bring an infusion of capital for energy conservation and historic rehabilitation upgrades and to realize the benefit of managing the properties as one project. CHT expects the sale of the properties to City Neighborhoods to be finalized in November 2010.

Property history: All of the properties are occupied residences located in the ONE neighborhood of the City of Burlington. ONE is the lowest income, most diverse, and most densely populated neighborhood in Vermont. Most of the structures are over 100 years old, and vary widely in architectural style and physical condition. The properties have been used as single- and multi-family residences since the late 1800s and early 1900s.

Project results: EPA Brownfields Assessment Grant funding was used to assess these 13 properties, which revealed that cleanup was not necessary except to mitigate lead and asbestos during property renovations. When renovation and construction activities are completed in spring 2011, all 13 properties will include energy-efficiency upgrades. Approximately 195 construction jobs will be leveraged for the project. During the development phase, a relocation plan will be implemented to ensure that any costs incurred by temporarily displaced residents will be covered by the project.

All properties, when renovated, will be used for affordable rental housing in perpetuity. While all of the apartments will be reserved for households earning less than 60 percent of median income, one-third will be further restricted to be affordable to households earning less than 50 percent of area median income. In addition, the CHT will retain the affordable housing restrictions in place for current tenants, most of whom are earning below 30 percent of area median income.

Project timeline

Fall 2009:	*Phase I assessments conducted*
Spring 2010:	*Development of project scope, design and cost estimate*
Summer 2010:	*CHT applies for project funding to Vermont Housing and Conservation Board, Vermont Housing Finance Agency for low-income housing tax credits and tax-exempt bond financing, and the Burlington HOME program*
Fall/Winter 2010:	*Construction begins*
Spring 2011:	*Construction complete July 2010.*

(Source: www.epa.gov/ne/brownfields/success/10/Burlington_VT_Old_North_End.pdf)

Environment and energy are vital components of community and economic development. It's impossible to have long-term resiliency without attention to these dimensions of community. As seen in this chapter, it requires collaborative efforts across sectors, with myriad partners. It's also about a mindset that both residents and policy makers hold – how are sustainable approaches valuing the environment viewed? And how do these values translate into actions?

Figure 6.2 Much of Burlington's housing stock dates from the early twentieth century and benefits from retrofitting
Photo: Bruce Seifer.

The Closer Look case selection centers on creation of the Vermont Sustainable Jobs Fund, bringing together the idea that development can be cognizant of the environment. This "green" approach to business development can yield powerful benefits. By the way, the nonprofit Vermont Sustainable Jobs Fund grew out of an idea recommended in the 1994 *Jobs & People III*. This organization is charged with accelerating the development of Vermont's green economy. Located in Montpelier, the state capitol, it was created by the Vermont Legislature in 1995. They provide early-stage grant funding, technical assistance, and loans to entrepreneurs, businesses, farmers, networks and others interested in developing jobs and markets in the green economy. It's interesting to note how early investments in, and commitment to, building social capital and environmentally sustainable approaches pays off long term. Nearly 20 years later, the benefits are being garnered.

A closer look: Vermont Sustainable Jobs Fund

"Vermont's green economy speeds up"[6,7]

In the summer of 1994, a group of about 15 business leaders convened for a day-long retreat in a rural meadow to discuss their long-term vision for the state of Vermont. All were members of an organization called Vermont Businesses for Social Responsibility (VBSR), an organization whose mission is to support and encourage socially responsible business practices and public policy initiatives. The leaders committed to a vision that included the idea of creating a state entity to support the development and creation of "sustainable" jobs, which were defined as jobs consistent with VBSR core values – protection of the environment, social justice, and economic equity (see "What Sustainable Means to Vermont," p.175). These leaders formed the Sustainable Jobs Coalition to pursue legislation that would translate their vision into reality.[8]

Since the 1960s, Vermont has had a reputation as a national leader on conservation, thanks to laws protecting the environment, including a landmark land use law, the billboard ban, the bottle deposit law, Green-Up Day, and the Scenic Preservation Council. Most companies that rely on Vermont's green image and fertile land to grow and sell their products approve, and many company leaders are in VBSR. They see an opportunity for growth through cooperation and networking, particularly in the area of marketing and advertising, where pooling resources can really pay off.

Consider one industry example of collaboration. Vermont's artisan and farmstead cheese makers rode a wave of national popularity in the 1990s. But as Cabot Creamery's director of marketing, Jed Davis, recalls, they were "really pretty darn insignificant compared to big states like California. We knew that we weren't going to gain as much by being competitors as we are by cooperating." To grow their businesses – and create new jobs for Vermonters – participants in the state's cheese industry knew they would need to work together to build their reputation as the premier source for small-batch, farmstead cheeses.

So VBSR developed the idea of a sustainable jobs fund. The fund would support growing enterprises and business networks that demonstrated commitment to a dual bottom line – making profits while pursuing social responsibility for the environment, social justice, economic equity, and an increased number of jobs.

The purpose of the Vermont Sustainable Jobs Fund (VSJF) was to create a nonprofit arm of the state that, with an initial state investment, would be able to attract funding from the federal government and private foundations to support the development of sustainable jobs. VBSR, in concert with allies such as those in the Sustainable Jobs Coalition, crafted and passed the legislation in 1995 with the modest appropriation of $250,000.

Grantmaking

The VSJF's initial approach was twofold: plant as many seeds as possible by providing grants, and build networks to support existing or emerging businesses. VSJF has acted as a catalyst, leveraging good ideas, technical know-how, and financial resources to propel innovation in sustainable development, especially in the realms of organic agriculture and local food systems, sustainable forestry, and biofuels (locally grown for local use).

Since 1997, the VSJF has made grants of more than $2.7 million to 150 recipients. Grantees have utilized these funds to leverage an additional $11.8 million to implement projects, test ideas, and assemble the building blocks of a green economy. Their combined efforts have created approximately 800 local jobs, supported community development initiatives, preserved resilient ecosystems, filled vital needs in Vermont's economy, and provided new models for moving forward.

VSJF funding in 1998 and 2003 was used to help form the Vermont Cheese Council (VCC) and to help its first 12 members market their products, share technical assistance, develop quality standards for Vermont cheeses using the Council's label, develop a fund-raising plan, complete the VCC website, and produce a logo. Today, two-thirds of VCC's now 36 members have won awards. At the 2006 American Cheese Society Conference, the Clothbound Cheddar collaboration between Jasper Hill Farm and Cabot Creamery won best in show.

"People were scratching their heads," laughs Davis.

> *Cabot makes the cheese? And Jasper Hill ages the cheese? And you don't fight about this? For that cheese to win best in show provided validation for everything that Vermont is all about. Not that there's just great cheese coming out of here, but that our whole approach to it is innovative and unique.*

The VCC represents a transition from commodity to value-added agricultural production, which yields a higher rate of return for farmers and a more diverse range of local products for consumers.

Network building

With limited funds, VSJF was forced to become innovative in supporting businesses. VSJF saw that every successful business is embedded in a network of relationships, and the stronger the network, the more sustainable, flexible, and resilient the business. The power of networks was a notion derived in part from Robert Putnam, author of *Bowling Alone*, who refers to social capital as "the connections among individuals (social networks) and the norms of reciprocity and trustworthiness that arise from them."

Social capital can be linked positively to innovation, to sales growth, return on investment, international expansion success, and the like. Given Vermont's size and connectedness, it's not surprising that the state is ranked third in the country according to Putnam's social capital index. And VSJF saw that creating a supportive environment to nurture and sustain these kinds of business networks and organizations would be a prudent way to use limited development resources.

VSJF grants helped enable more than a dozen networks, representing 1,600-plus businesses, to do the kind of strategic planning, capacity building, information sharing, market research, joint marketing, and policy development that are crucial to developing a unified voice and competitive advantage in a sustainable economy.

Market building

VSJF also recognizes the importance of developing markets for existing businesses. They take the perspective that markets are made through interactions among businesses, government, nonprofit organizations, communities, and other resources. For example, four years ago biodiesel was not available in Vermont. VSJF and its partners conducted pilot projects and educational activities that successfully introduced biodiesel to large-scale institutional and commercial diesel users, in addition to residential heating oil customers.

To realize this success, VSJF in cooperation with the University of Vermont Extension Service, conducted on-farm oilseed production and feasibility studies to help farmers familiarize themselves with these new crops. They helped a small nascent biodiesel producer expand his capacity with new equipment, enabled several farmers to construct on-farm production facilities, and assisted the installation of biodiesel pumps at a fueling station. With a little over $2 million invested over the past four years, more than 30 locations in the state now carry biodiesel, and many farmers are in the process of developing farm-scale biodiesel production capacity.

VSJF also has committed to extending its influence outside Vermont borders. It is now in the process of codifying Vermont's model of local production for local use into a set of sustainable biofuel principles, policies, and practices that could be applicable in other rural states.

VSJF's experimentation has affirmed that a little goes a long way. With investments and technical assistance targeted at the development of markets for sustainably produced goods and services, the building blocks of Vermont's green economy are ready to go mainstream.

What "sustainable" means to Vermont

The Vermont Sustainable Jobs Fund was established by the Vermont Legislature in 1995 to build markets within the following natural resource-based economic sectors:

- environmental technologies
- environmental equipment and services
- energy efficiency
- renewable energy
- pollution abatement
- specialty foods
- water and wastewater systems
- solid waste and recycling technologies
- wood products and other natural-resource-based or value-added industries
- sustainable agriculture
- existing businesses, including larger manufacturing companies striving to minimize their impact
- waste through environmentally sound products and processes.

The VSJF works with entrepreneurs and consumers to develop both the supply of and demand for goods and services that provide sustainable alternatives to economic practices that could cause negative impact over time.

Success has come through a combination of targeted, early-stage funding, technical assistance such as business coaching and the Peer to Peer Collaborative, and a focus on the long term.[9]

Resources and ideas for making it happen in your community

Burlington's Home Depot Sustainable Community Development award: see the video at: www.homedepotfoundation.org/video-burlington2010.html

The City's Brownfields Program has been the subject of four USEPA "Brownfields success stories." The links to these are:

Vermont Transit Bus Barns project on North Winooski Avenue: *www.epa.gov/region1/brownfields/success/burlington_vt.html*

Waterfront apartments on Lake Street: *www.epa.gov/region1/brownfields/success/06/waterfront_apts_burlington_vt_ag.html*

The Miller Community Center on Gosse Court: *www.epa.gov/region1/brownfields/success/09/R1_SS_Gosse_Armory_VT.pdf*

The "City Neighborhoods" project: *www.epa.gov/region1/brownfields/success/10/Burlington_VT_Old_North_End.pdf*

Climate Action Plan: *http://burlingtonclimateaction.com/*

Energy improvements, when to finance: When a business or organization is buying a building, conduct an energy audit. Incorporate the cost of the energy-efficiency improvements into the mortgage to match the term of the loan with the potential life of the energy-efficiency improvements financed. This can save significant amounts of money and energy over time.

Improving the landscape: Find an area that feels barren and plant trees, gardens, community gardens, and create a park. Nurture this space over time. Support your local Green-up Day where trash is collected throughout the city in early spring with volunteers. If you don't have a Green-up Day, start one.

Trade associations: Join trade associations to network and learn from your peers

Develop trade associations: Create new trade associations where none exist in your city, region or state to work cooperatively to solve problems, learn from each other and promote your collective interests. Incorporate fun into regular meetings to build team spirit.

Capital ideas: Train volunteers and staff on loan underwriting. Use and leverage local revolving loan funds. Many communities have underutilized loan funds that can leverage other resources to grow your economy.

7 Tasting as Good as It Looks: Local Food System Sustainability

How Is a Food System Sustainability Maintained in the Context of Community and Economic Development?

The average piece of food in the U.S. travels 1,500 miles from the farm where it is grown to the table where it is consumed.[1] This is a remarkable figure and one that holds all sorts of implications for communities and economic development – not the least of which centers on sustainability. Food security issues have come into the forefront too, not only in relation to supply and access but also quality and safety of our food supply.

All of these concerns and more are prompting a local foods movement in many places, with residents and visitors alike seeking out local sources as well as local experiences. Burlington is housed within a state that is noted for its rural working landscapes and focus on healthy food. Burlington was named as "healthiest city" by the Centers for Disease Control and Prevention in 2008 and Vermont was named the 2010 healthiest state in the U.S.[2] It's a natural fit that local food systems serve as a priority for community and economic development. Initiatives to increase awareness of the benefits of a local food system have been key, as well as concerted efforts across political boundaries to build support and collaborations.

Why food and development? Food systems, particularly sustainability and security aspects, represent a major component of an environmentally sustainable economy sector as well as social and inclusiveness components. This in turn helps foster the creation of a durable economy. As an essential component to life, food merits attention as a community and economic development focus. The United Nation's universal human rights framework, the Universal Declaration of Human Rights (U.N. 1948), Article 25, establishes the right to food as, "Everyone has the right to a standard of living adequate for the health and well-being of himself [sic.] and his family, including food." It is clear that food and food systems are central to individual and community well-being.

Hungry for more?

One can take a fascinating foray into the complexities and challenges of local versus industrial food issues with recent books such as *Farmer Jane: Women Changing the Way We Eat* by Temra Costa (2010) profiling 26 women in the sustainable food industry; *The Town that Food Saved: How One Community Found Vitality in Local Food* by Ben Hewitt (2009) chronicling the story of Hardwick, Vermont's efforts to build a vibrant food economy; or Vermonter Ron Krupp's compendium *Lifting the Yoke: Local Solutions to America's Farm and Food Crisis* (2009). There are classics too (both instant and long standing) that need reading so the foundations of the local food movement can be understood: *Diet for a Small Planet*, the 1975 classic by Frances Moore Lappe focusing on why there is hunger in a world of plenty, and the next generation of this, *Hope's Edge*, with daughter Anna Lappe (2003); Michael Pollan's *The Omnivore's Dilemma* (2006) illuminating America's "national eating disorder;" and *Fatal Harvest: The Tragedy of Industrial Agriculture*, Andrew Kimbrell's 2002 myth-dispelling work. The literature is rich in this area, these are just a few of our favorites!

Setting the table: local policy for supporting food systems

As noted earlier, one of the seven key recommendations of *Jobs & People IV* (2010) is that CEDO continue to focus on Local First type initiatives (see Chapter 2 for more details). This is not only about businesses producing goods and services locally and using locally sourced materials where possible, but also changing buying habits of people to consume these goods and services. Food products comprise a large part of this. This requires continued capacity building in food systems, as well as various marketing initiatives to entice more residents and visitors to source their foods and eating experiences locally.

Burlington's commitment to urban agriculture and value-added food production is strong; there are several projects relating to food and community development in progress, planned, or already completed. These include:

- Intervale Center (see case in this chapter under the Closer Look section).
- Fresh District (planned effort to promote a fresh foods corridor).
- Food Enterprise Center (an initiative for generating food-based businesses at the Intervale Center).
- Value Added Food Sector (plans to focus on seeking out opportunities for producing food products in Burlington).
- ONE World Market (see information on this in Chapter 3, several local food-related businesses were included).
- CEDO tasked in 2011 to help facilitate the development of a new comprehensive Urban Agriculture/Food Policy for Burlington. The completed Urban Agriculture Task Force Report was adopted by the City Council in September 2012.

Policies to support a local food system started long ago. The 1994 *Jobs & People III* established a food policy for the City. The initial goal was to locally produce 10 percent of the food consumed in Burlington and have at least 100 people employed producing this food. This 10 percent goal was first promoted by Will Raap, founder of the Gardener's Supply Company and initiator of the Intervale Center.

Burlington Urban Agriculture Task Force

Report to Burlington City Council

"Put your faith in the two inches of humus that will build under the trees every thousand years"

—Wendell Berry

September 2012

Authors: Alison Nihart, William Robb, and Jessica Hyman

With contributions from: Jimmy DeBiasi, Kyle Clark, Caryn Shield, Carly Schwer, and Rachel Cairns

Task Force Members: Ed Antczak (CEDO), Will Bennington (Farmer), David Casey (Board of Health), James DeBiasi (Intern), Jessica Hyman (Advisory), Alison Nihart (Researcher), William Robb (Chair), Harris Roen (Planning Commission)

Figure 7.1 The Urban Agriculture Task Force Report was delivered to the City Council in 2012

Just a few years ago, the Vermont Businesses for Social Responsibility (VBSR) public policy committee developed a plan for a "Farm to Plate" study. Funded with $150,000 from the State of Vermont, the study was presented to the Legislature in early 2011. Conducted by the Vermont Sustainable Jobs Fund (see Closer Look case in Chapter 6), the goal of the study is to show how the state can grow at least 10 percent of the food it consumes. This is expected to generate around 1,700 jobs in the state's economy.[3]

Ten best cities for the next decade

They're prosperous, innovative, and they'll generate plenty of jobs, too

By the editors of *Kiplinger's Personal Finance* magazine

Burlington's local-food movement perhaps best tells the story of how environmentalism drives much of the city's economic growth. Many shops and restaurants along Burlington's Church Street Marketplace, the famous pedestrian mall, serve up local goodies. A couple blocks over, the City Market/Onion River Co-Op, a community-owned grocery store, offers more than 1,000 Vermont products. (And atop the supermarket, generating 3 percent of the Co-Op's energy needs – enough electricity to power six Burlington homes – are 136 solar panels from groSolar, another Vermont-based company.) And the crown jewel for locavores: The Intervale Center is a nonprofit organization that has managed 350 acres of family-owned farmland in Burlington since 1988 and provides 10 percent of the town's food.

(*Kiplinger's Personal Finance* magazine, July 2010)

Supporting initiatives in the Burlington community

Burlington has a long history of food focused initiatives, including those generated by community organizations. Several decades ago, Jim Flint worked with Gardens for All, which later grew into the National Gardening Association (NGA). This association serves as an educational resource promoting renewal and sustainability of the "essential connection between people, plants, and the environment." They describe their vision as a plant-based educator to make available "materials, grants, and resources that speak to young minds, educators, youth and community organizations, and the general gardening public in five core areas; education, health and wellness, environmental stewardship, community development, and home gardening."[4] Their headquarters are located in Williston and, as Jim points out, they started as a community advocacy initiative, as a way to stimulate both positive environmental and social community development outcomes. This foundation is still evidenced in NGA's activities –

> *We have a long-standing commitment to community initiatives … [we] keep this philosophy alive by developing programs that help rebuild neighborhoods, instill community pride, build self-esteem, and green up urban environments. Community gardens provide a source of food, add aesthetic value, encourage physical activity, help preserve cultural identity and, most importantly, cultivate neighborhood relationships.*[5]

> *Gardens are a great way to address the health aspects of a community. They bring together people for a common purpose and with good design can encourage positive environmental and social outcomes.*
>
> (Jim Flint, former director, Friends of Burlington Gardens)

A note from social entrepreneur, Robert Egger

To Bruce Seifer and the ENTIRE city of Burlington, VT

OK, I'll admit it … I was an eye roller about Burlington.

Coming of age in the late 70's, I tended to be a tad dismissive of hippies in general, so the thought of an entire town of aging, cow-hugging, tie-died, stoners led by Bernie Sanders, the only Socialist Mayor in America, seemed like more of a caricature than a city to me … but that image went up in a puff of smoke when I got an early morning tour from Bruce Seifer, a champion for (and the navigator of) economic development in Burlington for about 25 years.

Exhibit #1 and the thing that really got my attention (besides their alternate currency project, Burlington Bread, the crazy-cool public sculpture contests, their innovative housing plans, the glorious use of their waterfront properties, and the wicked smart application of public-private financial partnerships that makes this city so economically righteous) is the Intervale Center (www.intervale.org). They took the old city dump, their "brownfield" and have turned it into one of the most amazing 365 acres on God's Green Earth. It includes a massive compost center, that processes about 20,000 tons of organic waste annually, "Healthy City," a nutrition education program – "where food comes from" project, an extremely successful Conservation Nursery and very soon they will break ground on a Food Enterprise Center, which will have a 20,000 sq. ft. food processing center and 20,000 sq. ft. greenhouse space, which will be heated by the steam produced by the neighboring wood processing plant. They already grow about 6 percent of the food consumed in Burlington, and their goal is to eventually grow a full 10 percent of the food their citizens' need ... while also creating the kind of business ownership opportunities that keep both folks invested and money reinvested in the city. Simply put … this is about vision, health, sustainable use of land, investing PUBLIC MONEY and maintaining respectful partnerships … and it totally, works.

Chalk one up for the hippies.

(December 14, 2006, Robert Egger's Piece of Mind blog)

The Legacy Project, the partnership centered on implementing Burlington's sustainability vision by the year 2030, has several initiatives at various stages (completed, on-going, or proposed) for encouraging food systems outcomes. Here's a summary of a few of these:[6]

1 *The Burlington Food Council is one of about 100 food policy councils (FPCs) in the United States. The purpose of an FPC is to have stakeholders from diverse food-related sectors come together to explore local and regional food systems operations and to make recommendations for improvement.*[7] *To help promote a healthy food system in the area, the Burlington Food Council is focused on building capacity via relationships and education (see sidebar for additional information).*

Burlington Food Council

Mission

The Burlington Food Council is an open community group exploring ways to ensure that Burlington creates and nurtures a healthy, equitable and sustainable food system for all members of the community.

Approach

To accomplish this mission the Burlington Food Council provides networking, partnership building, and educational opportunities around food issues, and provides strategic recommendations for decision makers. The BFC also works to serve as a model and source of innovation for the many groups involved in creating and nurturing a healthy, sustainable, and equitable food system for the City of Burlington.

Burlington Food Council goals

Goal 1 Build food knowledge and experience for Burlington children, their families, the wider community, and BFC members.

Goal 2 Build local food appreciation and access for Burlington children, their families and the wider community beyond the school day.

Goal 3 Build local food systems and establish stronger links between food producers – including gardeners – and school-age youth, their families and other community members.

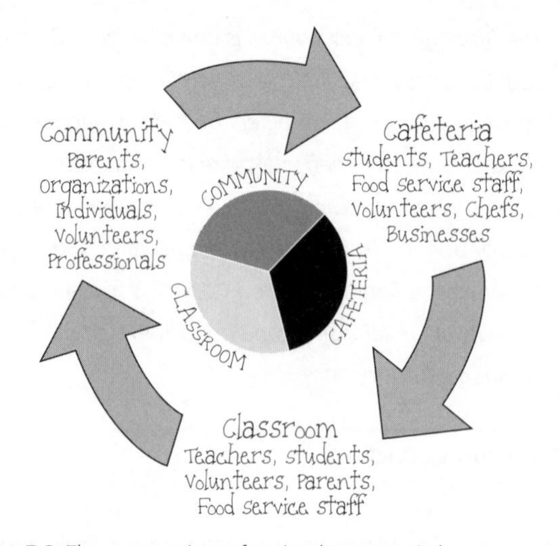

Figure 7.2 The community–cafeteria–classroom circle
Source: www.cedo.ci.burlington.vt.us/legacy/foodcouncil.html

2 *The Burlington School Food Project (BSFP), a partner initiative involving the School District, Sustainable School Program, City Market/Onion River Coop, VT FEED, Friends of Burlington Gardens/Healthy City Youth Initiative, Shelburne Farms' Sustainable Schools Project and others is a national model on how to bring local foods to school cafeterias, support farmers and improve childhood nutrition. It is Vermont's largest farm-to-school program and started with a USDA Community Food Project Grant in 2003. The BSFP recently received a $75,000 grant from Green Mountain Coffee Roasters to continue their efforts throughout the district's nine schools.*[8] *Its mission is to connect students and families with fresh, whole locally sourced foods with the intent of improving community health. The goals of the project are:*

(a) *Build the capacity for Burlington to better meet the food needs of students;*

(b) Increase awareness of and encourage healthy food choices for children and their families; and

(c) Improve Burlington's School District access to food choices from local farms.

They conduct the following activities to attain these goals: (1) incorporate whole, fresh, and local foods into school meals so that students have more opportunities to eat healthily and to practice healthy decision making; (2) educate students about food, farming, and nutrition so that they have the understanding to make healthy food choices; (3) extend these opportunities to students' families so that families can and will offer more opportunities to eat healthily at home; (4) help nurture the local food system so that whole, fresh, and local foods for school meals are provided; and (5) by doing so, the economic viability of local farms and farm-based businesses is improved, and creates opportunities for the Burlington community to make food decisions that enhance personal, economic, and ecological health.[9]

Figure 7.3 Burlington School Food Project garden
Photo: Friends of Burlington Gardens/Vermont Community Garden Network.

Growing our own: Fletcher Allen Health Care

Figure 7.4 Thomas Norcross, supervisor, Nutrition Services at Fletcher Allen Health Care, started keeping bees in the spring of 2009. As part of the hospital's effort to support local food sources and produce some of its own food, he started keeping bees for the institution in 2010. New hives in 2012 produced approximately 60 pounds of honey, a number he expects to rise significantly as the hives develop

Photo: Jordan Silverman. www.fletcherallen.org/about/environmental_leadership/sustainable_nutrition/

3 *The Harvest Cafe, opened by Fletcher Allen Health Care, Burlington's Medical Center/Hospital, incorporates fresh, organic, and local fare, such as local squash and soy milk, locally-raised ground beef, chicken and turkey raised without non-therapeutic antibiotics and arsenical compounds, vegetarian choices, and organic fair trade coffee. The goal of the cafe is to be the most sustainable retail establishment in health care, not just with seasonal foods offered year-round, but with environmentally friendly furnishings, such as nontoxic flooring, for example. Their refrigeration system uses outdoor cold air during the winter months for an energy-efficient "freeaire" system (the first of its kind in any Vermont hospital). This technology was developed by Richard Travers of Freeaire Refrigeration, located in Waitsfield, Vermont. As the State of Vermont's largest 'restaurant',*

serving 1,400,000 meals a year, they spend a quarter of their food budget, or $660,000, on the purchase of local foods from area farms and businesses. They source fish from North East Fisheries through Community Supported Fisheries (CSF) which is similar to Community Supported Agriculture.

Checking the almanac: looking to the future of food systems, community and economic development

So what does the future hold for food systems, community, and economic development? It isn't an easy task to become more self-sufficient, considering what local food producers face against the very large industrial food industry. Certainly, the outlook is sunny with increasing sales of organic foods. U.S. sales of organic food and beverages have grown from $1 billion in 1990 to $24.8 billion in 2009. And despite economic recession, sales in 2009 represented a 5.1 percent growth over 2008 sales – and the highest growth in sales during 2009 were organic fruits and vegetables, up 11.4 percent over 2008 sales.[10] This doesn't paint the entire picture, however. As more industrial agriculture "giants" enter the organic food production business, it changes the scenario for small, local producers. It becomes even more important to develop community capacity to support local producers, and to educate the public on the value of sourcing foods locally. With growing concerns over health impacts of low quality food, locally sourced food alternatives hold particular relevance. Coupled with demonstrated (and experienced firsthand) community impacts of local food systems, this will help strengthen positions of local producers, processors, and providers. It requires collaborative efforts and the ability to work across organizational, cultural, and political boundaries.

The Vermont Sustainable Jobs Fund (VSJF) in consultation with the Vermont Agriculture Council has written the comprehensive ten-year Farm to Plate Strategic Plan to strengthen Vermont's food system in 2011. The VSJF is assisting other New England states with their food system planning efforts, and the New England states are coordinating their implementation efforts to strengthen the regional economy.

Burlington has integrated local food systems into overall community and economic development action. Food security and sustainability are crucial issues that deserve much attention. One other dimension not yet discussed is the related outcome of agritourism and culinary tourism that have developed due to the capacity of the local and regional food systems. Both agritourism and culinary tourism have significant development impacts on the city and region. Overall, the tourism sector is a major source of local employment, growing faster than the national average. It's an "export" industry for Burlington due to drawing people from other places, impacting the local economy with their spending. Of course, agritourism and culinary tourism are not the only components of the tourism sector in Burlington, but it's safe to say there is an economic development impact contributed by these activities.

Additionally, in 2010 the University of Vermont instituted a transdisciplinary focus in food systems.

> *The Food Systems Spire focuses on the critical role of our local, regional, national and global food systems as they, in turn, affect soil and water quality, human health and nutrition, global economics, packaging and transportation interests, and overall food and energy security. Food Systems is a nascent field of study nationally and an emerging strength at UVM that is particularly well-suited to the UVM Land Grant mission of research and engagement in the 21st century.*[11]

The continued focus on existing efforts as well as expansion into new initiatives lays the groundwork for long-term food system sustainability. One of the seminal efforts is the Intervale Center, a remarkable bread basket production, educational and outreach center in Burlington. In the following Closer Look section is a summary of the story of the Intervale, a community food system hub. The name Intervale implies low-lying land by a river, which is exactly where it is located.

A closer look: the Intervale Center[12]

Figure 7.5 Intervale Center logo
Source: Intervale Center.

Twenty years ago, the Intervale, along Burlington's Winooski River, was literally a dump. With true vision and determination, a group led by Will Raap and Gardener's Supply Company came together to transform the Intervale into a thriving community of agricultural entrepreneurs and an unparalleled community asset. Since 1988, the nonprofit Intervale Center has pushed the limits of food system innovation, establishing itself as a leader in what is now a burgeoning global community food systems movement. Today, this 501(c)3 organization is working to fulfill its mission to strengthen community food systems, transforming our food system from one that relies heavily of fossil fuels and chemical inputs, destroys ecosystems and human communities, and devalues the contribution of producers, to one that protects land and water resources, contributes to healthy communities and economies, and honors the producers whose work quite literally sustains us.

What was once an informal community dumping ground is now a unique community resource – 350 acres of farm fields, wetlands, trails, and wildlife corridors that are home to 14 independent farm businesses and provide recreational and educational opportunities for hundreds of Burlington residents each year.

How did this remarkable transformation happen? Here's the story.

"Why not grow Vermont's fresh food in Vermont, and do it sustainably?" That was the question preoccupying Will Raap in the 1980s when he had a small garden shop and catalogue, a compost pile, and a parcel of neglected land in Burlington. Back then, the "Intervale" referred to 350 acres that were historically important but had fallen into disuse.

Based on consumer research, Raap saw in the land the potential to grow at least 10 percent of Burlington's fresh food at the Intervale. Today Raap believes that the gardens and farms on the Intervale have just about reached that goal. But equally important was how he got there. By creating a successful composting operation and inspiring the development of a dozen small farms for the city, Raap's initiative was able to revitalize the land and generate enough cash flow to support the nonprofit Intervale Center, a web of food-related enterprises and farm viability programs that have become the backbone of the northern Vermont's food system.

Figure 7.6 Fair held at Healthy City Youth Farm in Burlington's Intervale
Photo: Friends of Burlington Gardens/Vermont Community Garden Network.

Intervale at a glance

Where:	Burlington, Vermont
What:	Develops sustainable agricultural and community food systems solutions
Who:	Will Raap, founder
When:	Started 1988
Number of employees:	14
Total revenues:	$1.3 million

The Intervale area of Burlington now houses several businesses. In addition to the Gardener's Supply family of companies, the area includes Burlington Electric Department's McNeil Electric Generating Station (a 50-megawatt biomass powerplant), the Sugarsnap Cafe and the Stray Cat Flower Farm and Market. Linked with the Intervale Center itself are over a dozen independent farms that lease land from the Intervale Center. The Center is home to a multi-farm Community Supported Agriculture (CSA) program and distribution business, a conservation nursery, a farm business planning and consulting service, and a food hub. According to the Intervale Center's executive director, Travis Marcotte,

> *The Intervale Center's work includes all the working parts of a strong community food system: we have excellent agricultural land; we incubate farms; we provide distribution and marketing for farmers; we provide farm business development services across the state; and we foster a community of engaged and empowered consumers.*

How do these businesses help make Burlington more sustainable? On site, the McNeil Station generates most of Burlington's electrical power primarily from sustainably harvested Vermont wood chips. Intervale Compost Products (relocated to a neighboring town in 2011) transformed the city's organic waste streams into compost and topsoil sold commercially in and around the city. The Intervale Center stewards 350 acres of land and supports a number of programs that strengthen the food system. And all this business activity has actually helped revive the ecological vitality of the Intervale itself, such that multiple farm enterprises can produce more than a million dollars of organically grown food for local consumption each year.

Business model

The Intervale Center is a nonprofit that engages local farmers and consumers at every step of the supply chain of local food, from pre-production planning to post-consumer waste disposal. It has a farm enterprise business incubator for new farmers, business consulting services for established farmers, and a land preservation initiative. The state's first CSA is based there (it has since spun off as an independent consumer-owned cooperative).

As an organization comprised of multiple programs, initiatives, and enterprises, the Center is continually evaluating its operations against a triple bottom line of profitability, environmental sustainability, and social responsibility. One of the Intervale Center's signature accomplishments is blending for-profit management with nonprofit enterprise. Couple that with the Intervale Center's deep sense of place and its commitment to the surrounding land, and you've got the Center's operational philosophy. As Marcotte notes,

> *We're blessed to be stewards of a great community resource, which gives us opportunities to build programs that support*

farming, improve land and water quality, and benefit our community. This land – this place – is the heart of what we do and why we do it.

Unlike most nonprofits, the Intervale Center has placed a high premium on the financial sustainability of most of its programs. It works to provide 60 percent of its annual operating budget from the sale of goods and services and the rental of land and properties. The remaining 40 percent is raised through grants and community fundraising. An example is the Farms Program that has provided start-up support for emerging and small organic farmers since 1995. Incubator farms get subsidized rates and access to equipment and mentoring. The fees start low and only rise as the farmers' independent businesses expand over 3–5 years.

A newer Center program, Success on Farms, continues the focus on farm viability outside the Intervale. This program, funded by the Vermont Housing and Conservation Board, provides customized business planning and technical support services for growing farms throughout Vermont.

The diversity of its programs and impact make the Intervale Center unique. "What's interesting to me is that if you get the food system right, you get a lot right," says Marcotte.

If you get good healthy food to people, they are healthier. If you effectively steward land and protect water resources, the environment is healthier. When you enhance farm businesses, the economy is healthier. The Intervale Center touches many aspects of our community, economy, and environment everyday through the work that we do.

The Intervale Center's economic model leverages revenue from its most profitable programs to underwrite other start-ups or initiatives with stronger social missions. Programs such as the Intervale Food Hub and Intervale Conservation Nursery relied heavily on grants during their development and now generate revenue to partially or fully self-support their operations. New ventures often emerge from the direct needs of Intervale farms and the broader farm community, such as the identification and documentation of distribution and storage needs.

The desire to increase self-financing also has led the Intervale staff members to redesign or relocate programs that are no longer effective. For instance, the composting program outgrew the Intervale Center's ability to successfully manage it and was sold to the Chittenden County Solid Waste District. The Healthy City program, a celebrated youth agriculture program, was transformed from a difficult-to-fund youth farm project to an educational gardening program for Burlington's at-risk kids that is now housed at the nonprofit, Friends of Burlington Gardens, where it continues to receive acclaim in the national press as a model farm-to-school initiative.

Should the Intervale Center have been a for-profit? "I could argue either side of that," says Raap. "I do think that because we were talking about becoming stewards of a very large portion of open space and public land, we had to be a nonprofit entity. The City couldn't sell two hundred acres to a private buyer." But, Raap adds, had the Intervale Center been a for-profit, it might have attracted private finance and been able to move more quickly on some of its business ideas.

Several factors have contributed to the success of the Intervale Center. The model could not have happened had there not been a large and underutilized parcel of land. It also required a formal partnership with City, county, and state officials who provided the leverage to secure the land and needed capital. And it was critically important that, despite its nonprofit structure, the Intervale Center kept business development at the core. Its business model essentially targets the training and incubation sector; and a local market whose customers are local and regional farmers, food entrepreneurs, conservation agencies, local consumers, and vendors. It considers its niche as: organic, urban agriculture, land restoration and preservation, farm enterprise incubation, community outreach and youth education programs, with products including entrepreneur incubation, community programs, local economic development, local food system development, and agricultural marketing.

History and drivers

Raap is no ordinary businessman. He seems to blend Bill Gates' large-scale ambition with the small-scale values of E. F. Schumacher, economist and author of *Small is Beautiful*. In fact, after getting an MBA and an urban planning degree, Raap actually went to England to work with Schumacher.

In 1983, Raap founded the Gardener's Supply Company (GSC), in part because of his interest in food systems.

> *When I was a student in California, I was watching agricultural valley cities shift to suburban malls, and watching them lose their identity. I believed for a very long time that investor agriculture was going to burn itself out as soon as the oil burned out.*

Despite his commitment to small scale, Raap had no qualms about growing his business to catalyze big social changes. Today, GSC is a successful mail-order company for home garden products that employs 250 people and is one of the largest companies of its kind in the United States.

Back in 1985, after two years of growth, he moved his first store to five acres at the entrance of the Intervale, where part of an abandoned pig slaughterhouse was standing. Two years later he approached then-mayor Bernie Sanders (now a U.S. Senator) with his ideas for growing Burlington's local food system. Needing a new solution to the city's growing solid waste problem, Mayor Sanders liked Will's offer to move into the Intervale and set up a compost operation. The City leased land for operation to a division of GSC.

The Intervale Center is a work in progress. Sometimes programs work and sometimes they don't. For instance, the Intervale Center's composting program got to be too big, complex, and legally challenging. The decision was made to lease the land and facilities to the Chittenden County Solid Waste District. The District was able to hire existing staff and easily take over the operation because it had been a key partner in the operation from the outset.

Another ambitious project not quite realized has been the EcoPark. The idea was to create an "industrial ecology" model, where the waste of one business would provide the inputs to another. Will recruited John Todd, one of the pioneers of the concept. The excess heat from the City-owned McNeil power plant was to support a beer company, whose water waste and mash would feed into a greenhouse, where mushrooms and hydroponic vegetables would be grown and tilapia fish would be harvested. The political vagaries of federal funding and complications with a neighboring industry sidetracked the effort in 2002. The Food Enterprise Center emerged as a more modest and simpler concept, still incorporating the use of excess heat but focused more narrowly on season extension and value-added processing.

Raap himself co-founded another company that specializes in rangeland and farmland restoration using carbon credits, wetlands banking credits, and other payments for ecosystem services. His underlying philosophy – that the way to grow markets for local food is to increase the number of local farmers and the quantity of productive local farmland – remains the same, only now he is applying it outside of Vermont. He's developing the first organic CSA in Central America as one land-based enterprise in a 25,000-acre watershed restoration project. He still keeps one foot in

Vermont, however, and is developing a 20-acre organic community farm at the South Village Community Conservation development.

Key challenges and lessons

In its 23-year history, the Intervale Center has faced five big challenges:

- *Diversification* – For most of its history, the Intervale Center's cash flow was too dependent on its composting operation. A couple of years back, too much rain brought the operation, and its revenue, to a halt for several months. The Center has since worked to diversify its cash flow. Income from land leased to farm enterprises, along with the establishment of new enterprises such as the Intervale Food Hub and Intervale Conservation Nursery, has helped, but there is a need for even more diversification.
- *Focus* – Raap appreciates that to succeed, the Intervale Center must keep to its expertise in self-financing food systems. He thinks it was smart to hand over the compost operation to Chittenden County. "Just like in most businesses today, you've got to focus on what you're best at and get rid of the non-core initiatives."
- *Politics* – The initial relationship with the City was critical to the startup of the Intervale, but it made the Center appear closely aligned with Vermont's "progressive enclave" and subjected it to additional scrutiny. The result was a major regulatory dispute over the size and operation of the compost operation, once it grew to the point where it needed special licenses. It cost the Intervale Center $300,000 to come into compliance and effect the transfer of the operation to the solid waste district of the county.

- *Leadership* – Raap acknowledges that one of the key challenges for the Intervale Center has been himself. As long as GSC was providing funding to the Intervale Center, purchasing much of the compost, and leasing land to the organization, Will's ongoing presence on the board presented a lurking conflict of interest. The Intervale Center had reached the point where it needed a board and manager totally independent of GSC. However, his departure triggered challenges within the organization for about five years.
- *Next generation programs* – If the Intervale Center is to meet the demand for the local food it is promoting, there will need to be more farmers and growers in the region, and the Intervale Center needs to target some of its incubation work accordingly. Increasingly, the tactic for the Intervale Center is to use a systems approach. The Intervale Food Hub and its first enterprise – an extension of a CSA model that involves multiple growers – is a good example of how the Intervale Center has moved deliberately toward making systems thinking a high priority. Similarly, its commitment to land access for farmers beyond the Intervale demonstrates that it understands its land constraints.

Despite these challenges, people now come from all over the world to visit, study, and replicate its work – so many, in fact, that the Center has established a more formal consulting arm. With the Center committing to extend its mission and work beyond Burlington and Vermont, Raap felt comfortable enough to move onto other projects.

"I think the Intervale Center," Raap reflects,

> *has charted a way to stop food production from moving farther away from Burlington and from reversing the decline*

in the percent of food retail dollars going to our farmers. The food system impacts we have had over 20 years are a main reason Burlington is in fact recognized as among the most sustainable cities in the United States.

Resources and ideas for making it happen in your community

Burlington Food Council: http://burlingtonfoodcouncil.org/

Burlington Legacy Project: http://burlingtonlegacyproject.org/

The Intervale Center: www.intervale.org/

National Gardening Center: http://assoc.garden.org/

Food policy: Create a local food policy by working with the people interested in the subject. Include goals for increased food grown and consumed locally, additional jobs created and ordinances that could be changed to support this effort.

8 Summing Up

Lessons Learned and Other Insights

Many issues remain to be tackled, and there are no easy or quick fixes for any of these, especially conditions such as pervasive poverty and high costs of living. There have been many successes as well as ideas, programs, and projects that didn't work. While Burlington is doing well, there are still many strides to be made (for examples, see the Legacy Plan's 2010 Report Card, link provided in the Resources section of this chapter). Progress evaluation is part of the process and helps to show where Burlington has been and set direction for where it needs to go for the next 20 years. Progress evaluation is essential as it helps identify and redirect strategies and approaches that aren't working, as well as those that do work.

The final Closer Look selection is emblematic of what it takes to do community and economic development as discussed in this book. The case chronicles the long and tedious process of developing City Market, a downtown grocery cooperative, on a brownfield redevelopment site. It illustrates the complexity such an undertaking entails; the need for establishing an effective, participatory process; and the long-term commitment and laser-like focus on the desired outcome that is required.

Final analysis? It boils down to the question asked in the beginning of this book: What do residents want? Revisiting this question each time a choice is presented keeps the focus and gauges success. It also helps to think about it in terms of creating good processes, forging collaborative relationships, and producing results for the community. Creating a durable economy requires great effort from the public, private, and nonprofit organizations in a community and brings together many components: cultural, political, moral, social, environmentally sustainable, and locally-focused economy sectors. Just addressing a few of these won't work, but bringing them together and ensuring that these dimensions are incorporated can build a more resilient economy that endures over time.

Burlington has weathered the economic downturn well. Commercial development projects are increasing. There is a pattern to this type of economic activity – it's locally-owned companies helping stabilize and grow the economy. These are the types of activities that Burlington has long encouraged. It is not coincidental to have a strong employment situation in a national recession – organizations and companies with roots in a community tend to stay and invest in it.

Building capacity and willingness to take on challenges is a central feature of Burlington's approach. While each community has its own unique circumstances and issues, there are some commonalities, insights and perspectives to which everyone can relate. The next section presents insights from some of the community leaders in Burlington, followed by a summary of some of the lessons learned.

Voices from the community

Community partners are the reason Burlington has been able to achieve desirable community and economic development outcomes. They represent all sectors, not the least of which is the fourth sector, which are organizations integrating social purpose with business methods as described in an earlier chapter. The authors visited with many of them recently to gather their perspectives on what has and hasn't worked over the years. They provided insight into key elements to share with other communities in their efforts to build a durable local economy. In their own words, here is an opportunity to learn from them.

Peter Clavelle, first CEDO director and former mayor, City of Burlington

1. *Vision is important but you have to create the capacity to implement that vision. You have to build something like CEDO.*
2. *There must be an integrated approach to economic development. You have to connect all your activities, and an organization like CEDO provides integration.*
3. *You have to secure buy-ins from other stakeholders, it cannot be city government alone (government's job is not to row, but rather steer the boat).*
4. *Development created with principles with a priority given to sustainable development.*
5. *A sustainable community needs to stand on the four E's: economic development, education, equity, and environment.*

Doug Hoffer, Vermont state auditor and former CEDO staffer

Here's the reality of anything in a community: You can't do any of this without electoral politics – without it, nothing will happen. If you want something, get it in the political process with meaningful discourse and reflected in regulations, ordinances, and policies, in writing.

Julie A. Davis, cofounder, Vermont HITEC

People are the most important component of the success of any project. The single greatest obstacle to growing jobs in our current economy is making sure people have both the academic and hands-on skills to perform. Our economies will grow and our community health will prosper when our education system embraces new models of learning and focuses on educating those most in need.

Pat Robins, founder, Symquest, and helped implement the Church Street Marketplace

You have to believe that you can turn around a community – build consensus or common sets of interest and not pit forces at each others' throats. If you want to get a town to work, you have to take care of it. Social services are a huge part of this, and business needs to understand. It's not just a bunch of restaurants, a mix is needed and social services are part of it. Communities need veteran and homeless shelters, and a string of services to support them. Church Street Marketplace is a commercial entity in the midst of urban challenges; we coordinated with United Way to get social workers to stay on Church Street, and these programs do work. Finally, keep retail strong. If retail struggles, balance is lost so keep it busy and strong.

Beth Sachs, founding director, Vermont Energy Investment Corporation

My advice? Have smart people run the city, envision what you want your community to look like and then energize others to get there. Be even bolder than us in Burlington.

Paul Bruhn, director, Preservation Trust of Vermont

Communities have to try and build their local capacities. Providing basic essential services is crucial to success too. It's also about charitable capital, community investments and social support to foster success.

Mark Stephensen, founder, Vermont Energy Contracting and Supply

There is some sort of spark that exists, something that creates life, and you have to create this "attraction" for people to come; they are attracted because it is an interesting and culturally diverse community. It is also critical to maintain families and schools. It's vital to protect our school support systems, as school health is paramount to community growth and sustainability. What percentage of taxpayers have kids in school? This is a critical issue as there's often a disconnect and this influences balanced and sustainable development.

Jim Lampman, founder, Lake Champlain Chocolates

Having a liaison between the city government and business is important, to stay connected but not getting in the way. Someone that communicates the needs of business owners well is important. It's good to have an incubator mindset and resources for generating businesses as this will attract other activities such as restaurants, small stores, etc., to help an area flourish. Finally, a long-term look to the future is vital, asking such questions as: How to keep up the city? Where are the sources of revenue (citizens, businesses, federal government)?

Michael Monte, former CEDO director and chief operating and finance officer, Champlain Housing Trust

1. *Focus on your internal strengths, stay local and stay focused.*
2. *Serve your people first by building facilities for yourselves and others will come.*
3. *Be like a gardener for small business: plenty of soil, water, nutrients, help weed it out. Focus on small and allow to grow to be major employers and economic players.*
4. *Support the creative class. Yes, anything is OK.*
5. *Organize groups so that all have voices!*
6. *Don't separate economic and community development foundations and functions, they are intricately integrated.*

Finally, a general rule is if it's a good place to live, then economic development will happen.

Brian Pine, CEDO, assistant director of Community Development Housing and Neighborhood Revitalization

Lessons learned? (1) Nurture the nonprofit sector, (2) Recycle scarce resources through the land trust model, for example, and (3) Create an environment for discourse and participation to share the future with citizens. Partner with the nonprofit sector for accomplishing housing, they are by far the best at carrying out the housing agenda. They are much more participatory than others and a nonprofit delivery system is the way to go. About the role of municipality in community and economic development – we have tried to focus on guiding principles, they are vitally important on how to pursue work and respond to opportunity. It gives a clear perspective of what we are trying to accomplish in support of guiding principles. CEDO has been successful in evolving and meeting the changing needs of the community while holding true to these guiding principles. CEDO has a strong commitment to an economic and social mission and has been able to help create lasting change to outlive any particular project or program. It's this ability to create lasting change that's made it work.

Yiota Ahladas, former CEDO assistant director for the Center for Community and Neighborhood

Here are aspects to consider:

- Empower all stakeholders and help them develop capacity to come together with a shared commitment to finding what works.

- Invest in process and develop capacity to dialogue with partners; the future can be discovered in context with partners.
- Relationships establish precedent to collectively imagine the future and solve issues.
- By articulating shared values, people feel empowered to make things happen and become unstoppable.
- Educate people about complexities of problems then listen to their responses.
- Intentionally develop expectations that city government is working.
- Foster nonprofit, grassroots and citizen leadership with forums, business networks, university partnerships.

Leigh Steele, former program coordinator of Step Up for Women, now with Vermont Works for Women

Collaboration is key, so much happens because we work together with the state, other nonprofits, and the City. Get the community involved! Focus on neighborhoods and sense of community, and build a strong environment for families. There should be lots of opportunities for families, and close the gaps by serving people who are lost now so they are included too in the social and cultural fabric of the community.

Melinda Moulton, CEO and redeveloper of Main Street Landing Company

Identify your community leaders to help guide your city: without a very strong group of people this would never have happened in Burlington (or anywhere else). Then plan and make it happen – a collaboration of human spirit with willingness and courage is necessary to protect the environment and protect the great vision that is needed.

Ben Cohen, founder, Ben and Jerry's Homemade

A short-term horizon perspective abuses and exploits community. Long-term horizon thinking is looking for benefits of generations to come. Success depends on maintaining community essence and a sense of belonging to a community. Economic diversity has to be with resiliency where sectors are diverse, with a focus on small business rather than the boom and bust cycles of a few large industries.

Hal Colston, founder, Good News Garage

Collaboration and having good partners is the key! It really comes down to relationships; how do we get to understand and get to know someone who is different from me? Asking people what they need, not what you think they need. The dominant culture needs to be inclusive, with meaningful relationships.

Jerry Greenfield, founder, Ben and Jerry's Homemade

Elect progressive government; some believe government shouldn't be active, but you can't just cut costs and lower taxes. Active governance can be cumbersome with lots of process, but it's worth it for long-term success. Invest in your community!

Clem Nilan, former general manager, City Market Onion River Co-op

Each community has its own unique characteristics, but before starting a major project like City Market, conduct a site and market study. It's important to have a good model too, one that builds on strengths of management. Local is esteemed now, even more than organic and reflects a higher value set.

Steve Conant, owner, Conant Metal and Light

Identify and recognize your community's core assets. Then protect and enhance them, being actively engaged in developing them too. Branding and marketing is helpful to spread the word about assets. A question to consider is, How to keep landowners motivated by something other than money? Shifting the focus to core values, having supportive zoning and other regulations in place helps protect a district's assets. Embrace opportunity so community can benefit from it.

Betsy Ferries, former director, Mercy Connections

I am a believer in being inspired by best practices. But, the question is then like with any change, you have to go back to your community and see who the allies are. Identify a small group of people who will make something happen. You can find an idea you like, but you have to do a needs assessment first – what is the big underlying problem? I believe women are naturals at organizing and running their own businesses and their flavor is an important part of community. This combined with the other roles women have gives them flexibility; the female element is significant!

Tom Longstreth, director, ReSource

It's important to create walkable and pedestrian-friendly areas in a community. Building on natural resources is crucial too to support an environment where people can run, skate, walk, bike, and be engaged in the outdoors. The City has done a good job of using community resident groups, of trying to involve people, and this has helped create continuity over time. Capacity building is important too, for example, the City's AmeriCorps program has created entry-level jobs for young, educated people. By bringing young people to the city, both vitality and capacity is enhanced.

Rachel Hooper, owner, The Bobbin Slow Fashion + Sustainable Design

The key element is going to your community and asking what they want! Throw ideas out at an economic development summit. Participate and start with some grassroots organizations. Every resource is very different and encourages people to be active. Local living economies are important to gain power in the local economy. Give local businesses the same benefits as big boxes receive to incubate more small businesses. How about a local stock exchange or local currency project?

Diana Carminati, former CEDO director and staff for Women's Small Business program

Look at what people want and need. There is a need for an appropriate process to help them think it through – it starts with who needs to be at the table. Working with people to understand the community-based economic development process is important. Understand and map a clear process. Creative financing, governance, all these elements are important, but the process must be figured out first. If you do community and economic development, you must start with the process! Then the rest can happen: feasibility, zoning, etc. Who will be the champion for a project? Who else is there? Find them and ask them to walk beside you.

Bill Truex, architect and champion of downtown Burlington revitalization and Church Street Marketplace

Our efforts started in the early 1960s, we needed to connect downtown with urban renewal. My advice? Have a strong planning process in place, with an excellent review team. Ensure that design is user friendly, and create a public "stage" as a focal point. For large projects, implement a clear management plan with professional oversight. Church Street Marketplace's success is due in large part to having excellent management and development oversight.

Martha Whitney, former CEDO staff for Women's Economic Opportunity Project

1. *Find and build progressive models where people have a voice.*
2. *Set a vision that leads the community.*
3. *Don't be intimidated by pushing the edge.*
4. *Ask who are the people in need and how do we meet those people's needs? Bridge low-income needs with the larger community's needs. Responding to this will require collaborations to initiate ideas and build leadership to implement a progressive vision.*
5. *Seek out impassioned and skilled people to get things done.*

Alan Newman, founder, Magic Hat Brewing Company and Performing Arts Center, and serial entrepreneur who helped start Gardeners' Supply Company and Seventh Generation

Create accessible urban space – cities I like have good public spaces. Plus, vibrant urban cores make things happen in a community. Create a strong vision of what the community is trying to accomplish and a city government that serves people – not regulation-oriented but people-oriented. Focus on single home runs, not grand slams.

Will Raap, founder, Gardener's Supply Company

The key is a process that inspires, energizes and engages all members in a community. Create a place where conversations about even the basics (potholes, for example) to the best ways to move the whole system forward can occur. Is your community subordinate to the global economy or can you find ways to increase capacity to help in the face of global economic challenges? Understand resilience – what is that and how do you get there in your community? Unlocking capacity is critical to achieving this. Capacity is impactful and enhances smart community and economic development; find and use untapped capacity in your communities.

Lessons learned: moving communities into the twenty-first century

Focus on people

Focusing on people is one of the most important lessons learned by the City of Burlington in its community and economic development efforts. Keeping people employed in quality job opportunities is an essential component of a durable economy. While unemployment rates have soared across the country, Burlington's economy has fared well. In September 2012, the U.S. Bureau of Labor Statistics announced that the Burlington metropolitan statistical area has the lowest unemployment rate east of the Mississippi River and is in the top 2 percent for all regions in the United States. While the September 2012 nation's unemployment rate was 7.9 percent Burlington's rate was 4.0 percent.

Burlington was able to keep unemployment rates low in part by listening to employers, workforce developments experts and the unemployed workers. CEDO found that there is a widening gap between position openings in various sectors of the economy and the skills of the unemployed. CEDO worked to close that gap by linking employers with applicable workforce training programs. For example, a local web development company expanded from 40 to 675 employees in seven years aided by the use of a model workforce-training program. This model program was conceived by a local nonprofit organization called the Vermont Healthcare and Information Technology Education Center.[1]

The Vermont HITEC model uses grant funds to provide training at no cost to unemployed and underemployed participants. The participants are immersed in a rigorous, full-time training program that leads directly to employment. The curriculum is designed by reverse engineering the job to be performed and includes both academic and hands-on training. Participants are guaranteed a position with the participating company if they can demonstrate their ability to perform in the position before they graduate. The model has successfully trained for positions in IT, healthcare, and advanced manufacturing and is applicable to many other industries as well.

Taking care of business

Taking care of business entities in the community must be a major priority if a municipality wants to build a durable economy. Business owners of all kinds need to be considered as key partners in advancing public policy initiatives. Identifying employer needs and creating or encouraging entities to meet those needs is essential. Burlington's CEDO opened communication with business by offering services including technical assistance to small businesses, a revolving loan program, and targeted networking opportunities and conferences. They then proactively worked to help make individual companies successful. For example, a local chocolate company came to CEDO looking for a larger space. CEDO identified a building next to an elementary school, provided financing and linked them up with other services including energy-efficiency programs and lean manufacturing with the manufacturing extension center. This company has continued to grow and has now purchased another building across the street.

Success breeds success

Successful projects build upon themselves. In the early 1980s Burlington was filled with vacant commercial buildings. CEDO invested in redeveloping vacant sites in each area of the city. These investments encouraged private investors to redevelop other sites in the same area, attracting business and further investment.

For example, CEDO invested in the redevelopment of an old industrial building located on a busy commercial street on the outskirts of the center of town. The building became a magnet for small businesses looking for incubator space. With support from CEDO, a business association was formed whose mission was to promote the assets of the commercial district. These assets included an active artist community who used the underdeveloped sites in the area for inexpensive studio space. This association, again with the help of CEDO, worked to promote the area with annual events that brought tens of thousands of visitors to the area each year. The district has subsequently become a thriving arts and business district. Currently, there are very few, if any, vacant commercial sites left to develop in the City of Burlington.

Nonprofits play a dual role

Nonprofit organizations often have missions in alignment with City policy goals. They are important both as employers and as potential partners for City-initiated projects. Investing in or partnering with nonprofits can provide benefits far beyond initial expectations. For example, Vermont Energy Investment Corporation (VEIC) began more than 20 years ago with a small staff and funding from CEDO and other community partners. After 11 years of providing ongoing technical support to customers interested in making their homes and businesses more energy efficient, VEIC won a state contract for delivering statewide energy-efficiency services. VEIC now employs nearly 275 people who have, since 2000, saved the equivalent of 120 megawatts of electricity consumption, and who provide consulting services all over the world.

Long-term vision

Some of the most important economic development strategies require long-term vision. Develop programs and projects which are based on future trends and that will have a long-term impact. Think of how these projects will affect your grandchildren. Energy efficiency is important for keeping business costs down. The return on investment for these kinds of project can take longer to recoup but can have huge payoffs. Burlington is fortunate to have a publicly owned utility that invests in energy-efficiency programs. These programs have saved ratepayers over $10 million annually, and have kept electricity demand flat for 20 years with electric rates lower than in most New England cities.

Stay focused on present needs

Success is often counted by the number of ground breakings and ribbon cuttings, but by the time these have occurred the hard work has long since been done and there usually follows many years of much less interesting activity. It may not be interesting or newsworthy to lend advice to a small business but it can produce astounding results over time. Stay focused on the present need and let the future publicity take care of itself.

Keeping pace with technology

Fiber optics, broadband and cell tower reception are now considered state-of-the-art essentials among the business community. Fifteen years ago access to such amenities was considered a luxury item. A robust, adaptable technology infrastructure is critical to attract and retain businesses. CEDO supported the development of a state-of-the-art, City-owned telecommunications company called Burlington Telecom. CEDO organized meetings with businesses and nonprofits to determine their needs to ensure these needs were met when the system was built. CEDO wanted to provide information to these organizations so they could adjust their business plans accordingly. Although fraught with controversy in a highly politicized environment, the presence of the low-cost, state-of-the-art technology has been essential to attracting and retaining business in the high tech industry. Burlington Telecom is now offering fiber optic symmetrical broadband speeds up to 1 gigabyte to homeowners and businesses.

Change is the name of the game

Making decisions based on policies that are good for everyone in the community will always be of benefit in the long term. Energy efficiency is now rarely questioned, while in 1990 there was considerable opposition. When in 1985 recycling 25 percent of Burlington's solid waste seemed farfetched, today the City recycles almost twice that amount.

Slow simmer

Community economic development projects can take a long time to develop. The redevelopment of Burlington's waterfront has been ongoing since 1985 and is likely to continue over the next 20 years. Many projects have very long gestation periods due to public process, permitting, land assembly, securing financing, public controversy, law suits, federal and state regulations, personnel changes at all levels, market forces, industry standards shifting, politics, and media controversy.

Do the homework

Market research can be an effective investment tool for economic development projects and should be a prerequisite for larger projects. For example, the last supermarket that served Burlington's downtown was departing for the suburbs and the City wanted to replace the existing supermarket. Until a market feasibility study was conducted, supermarket operators were not interested in working with the City to explore options for the development of a supermarket on City-owned land. Once the market feasibility study was complete, the City was able to attract and eventually select a supermarket operator who shared the City's goals.

Think local

Many communities have focused their economic activities on luring large companies with tax incentives or other financial benefits. Quick fixes such a luring a large company to town with expensive tax breaks rarely last and often can't address long-term needs and issues. These companies are not likely to be as committed as those who grew from within the community. CEDO has instead focused on investing in the infrastructure and creating networking opportunities needed for companies to grow. For example, Pine Street – once a downtrodden industrial corridor – is now a thriving art and entrepreneurial sector with significant locally-owned manufacturing and their ancillary retail presence. CEDO worked with a local college to locate their Emergent Media Center in the South End of Burlington. They are now providing interns and training to local businesses. Also, the graduates of this college, who are now being regularly exposed to this business district, are more likely to think of starting their own businesses in the incubator spaces of this district. The nexus of the college and incubator infrastructure helps to make both the education and the business sector successful and promotes the growth of local businesses.

Leverage community leaders

Every community has a pool of community leaders who have the capacity to support community and economic development activities. Engaging your community leaders will bring success well beyond your initial expectations. City councilors are a good source of leadership and have much to gain in supporting projects. For example, one city councilor was instrumental in building support for what is known as the Neighborhood Commercial Zoning district. This zoning designation was designed to protect the quality of living in the neighborhood while at the same time allowing for commercial development. This same councilor also helped to create a tree-planting program, reduced the number of bars allowed in zoning requirements, helped to create a citywide Kid's Day, and encouraged and monitored dozens of grants to homeowners and businesses. All of these activities played a role in making Burlington an attractive place to live, work, and play.

Keep up with the times

The internet has changed the way everyone does business. Social media, fab labs, hackerspace, tech shops; keeping current with the latest business practices and trends keeps economic development efforts relevant and can also help to create new business opportunities. A website company that began with five employees has blossomed to employing over 675 people. Needing to attract young college graduates with technical skills, it wanted to stay in the South End of Burlington because it was considered "cool" by the talent they wanted to attract. In another example, savvy young entrepreneurs in Burlington took the lead on establishing social media marketing events for small business owners to learn from one another about how best to use the internet for marketing. This is an instance where the community culture inspired individuals to help themselves without City investment and support. It is hard to imagine that either of these examples would occur in a town with vacant buildings and high unemployment rates.

Share information

Sharing knowledge of ideas and resources with entrepreneurs, other community and economic development officials and education providers quickens the pace of development. CEDO regularly convened conferences to share information and published several resource guides to distribute in various ways. For example, CEDO published a 200-page resource guide describing the 150-plus economic development organizations in Vermont. This guide was to simplify and expedite the flow of resources to the public and to the agencies that provide similar services. For example, if someone wants to be trained for a new career, this guide is the only publication in Vermont that lets you know what organizations provide education and training.

Leverage resources

CEDO has been able to garner tens of millions of dollars a year from the public and private sector to help support the community and economic development efforts in Burlington detailed in this book. At the same time, local investments and contributions are a key component of successful process and projects. Local contributions of time and money allow residents and business owners to participate in the development process, and foster a sense of vesting and value. Many community economic development projects requires a mixture of funding sources from the government and the private sector to develop the project, whether it is establishing a workforce training program, constructing a building complex in downtown, or running a business revolving loan fund. Take advantage of your resources, mix and match them: think of them as ingredients to create a community economic development "meal" you want for your community.

Involve everyone

Collaborating, communicating and networking are the key processes necessary for success. Network and engage all community members even when it initially doesn't seem to make sense. CEDO established an ongoing structure to process significant community and economic development activities called the Neighborhood Planning Assemblies (NPA). These neighborhood forums provide residents with easy access to information and opportunities for feedback. Their presence is very important in moving community projects forward or stopping them altogether.

Government works best as a catalyst and facilitator. You know you have done your job well when an issue is taken out of your hands and is being solved by the community. For example, software developers expressed concern that they were being ignored by state government in terms of attracting resources to their industry. CEDO helped to establish an alliance of software and technology companies that eventually was able to advocate for their needs on their own with decreasing levels of City assistance over time.

Develop leadership and nurture local businesses

Communities need to plan long-term to develop leaders, retain companies when an owner is retiring, and to start and grow new businesses to replace those that go out of business. CEDO developed leaders by running an AmeriCorps*VISTA program where CEDO trained over 400 members to be effective in their positions. Seventy percent of these people were hired at dozens of community-based agencies when their year of service was complete. CEDO also nurtured support for the Vermont Employee Ownership Center that helped 14 businesses to successfully implement employee-ownership succession plans. CEDO also established an ongoing business assistance and revolving loan program as part of an environment that encourages businesses to start and grow. CEDO loaned over $10 million to 140 businesses over the past 25 years.

A closer look: City Market–Onion River Co-op

Figure 8.1 City Market–Onion River Co-op, Burlington's downtown supermarket
Photo: Bruce Seifer.

City Market is Burlington's downtown grocery co-operative that opened for business in 2002. The market is an award-winning urban grocery, with a strong, community-focused mission. After facing significant opposition the co-op model of the City Market is now wildly popular – so much so that it is listed as "a place to go" in visitors' guides to Burlington. It has become an anchor food store for the whole downtown shopping area and strengthens the entire economic base of Burlington by attracting and retaining 1,500,000 shoppers annually. It has also served as the inspiration for the development of downtown food co-operatives in New Haven, Connecticut and Burlington, North Carolina.

The need to develop a supermarket downtown came up in the early 1980s when the only downtown grocery store, which was 10,000 square foot in size, announced that it would be closing its doors. The downtown grocery renewed their lease twice but eventually moved to a new 100,000-square-foot store in the outskirts of town, leaving downtown Burlington without a supermarket. With the nearest grocery store located more than a mile away, low-income residents without transportation would have been significantly affected by the change. The story behind the development of City Market is a study of perseverance and determination to locate a healthy foods option in the downtown area. It took 17 years from the initial idea to the opening of the store and the realization of this goal. The process illustrates the combination of politics, power, and persuasion.

Here's the chronicle of the story, as told by Bruce Seifer, former CEDO assistant director of economic development for 28 years, and founding Board member of the Onion River Co-op.

Connecting to community with a community-owned food co-operative

City Market, formerly the Onion River Co-operative, opened downtown in early 2002 welcoming 2,500 visitors the first day. The 32,000-square foot market now serves more than 4,200 customers daily. The Co-op had 2,000 members before it became City Market and now has close to 9,000 members, serving over 4,000 customers daily. The market works with over 500 Vermont vendors to feature the widest selection of local products in the state. Of the 191 employees, 75 percent are full-time and 65 percent live in Burlington. Average wages exceed livable wages in Burlington by $0.93 cents per hour, and on average, employees earn 25 percent more than those at conventional supermarkets. They also have a generous benefit package including 100 percent of premiums paid on health care insurance for full-time employees.

Phase I: historical context

During the last decades of the twentieth century, eight grocery stores serving the Burlington region closed, including three serving the downtown and Old North End neighborhoods of Burlington. That left just one market downtown, also rumored to be at risk of closure. The neighborhood has a lot of low-income residents, students, working families, empty nesters, elderly, and middle- and high-income customers along with over 10,000 employees working at area employers. With the nearest grocery store located more than a mile away, low-income residents without transportation would have been significantly affected by the change. The only other option was a local co-operative located just outside the perimeter of town. Many residents were unfamiliar with the concept of a co-operative and were disinclined to shop there for cultural reasons. CEDO made the retention and/or development of a downtown grocery store a priority. It took 17 years from the initial commitment of developing a downtown grocery store to the opening of the market in 2002. Initially, we did not envision that this market would include the co-operative concept.

There were three separate attempts to develop a supermarket. The first attempt involved a Request for Proposal (RFP) that we sent out to local developers to develop a supermarket on city-owned land. This RFP process was unsuccessful due to a lack of interest from developers. The second attempt was in 1986 when we applied for and were successful in securing a $1.6 million HUD Urban Development Action Grant (UDAG) in 1987 for the development of a grocery store with rooftop parking at the site. Initially, we were told by HUD staff that the combination of structured rooftop parking along with this type of store was not financially feasible. We were eventually able to persuade HUD staff that it would work by going through the financial projections and the UDAG was approved. We then became intent on finding a developer, getting the permits and securing an operator to lease the site.

The project had interest from several developers and CEDO eventually chose to work with a new local development firm. We had spent three years getting the funding lined up, the developer on board, the permits in place, and the Request for Proposal approved. Unfortunately, by the time we had the project ready for construction, no grocery store was interested in leasing the 24,000-square-foot building. The lack of interest was due in part to a change in the development patterns. We were faced with the beginning of the era of building big box retail supermarkets in suburban locations. Downtowns were being boarded up across America and Burlington was struggling to maintain essential services and remain vibrant. The grocery store industry had changed from using a minimum of 24,000 square feet to nothing smaller than 45,000 square feet, making the project unpalatable to potential supermarket operators. In 1989, we had to return the $1.6 million UDAG to HUD due to a lack of interest from supermarket operators, and go back to the drawing board.

Phase II: finding a permanent solution

Marketing studies and site selection

The third attempt involved a more comprehensive approach. Learning from our past experience, our new plan to develop a supermarket involved matching a supermarket operator with the most appropriate site plan – not necessarily the City-owned site. The plan was to include a market analysis, public forums, a consumer preference survey, and a Supermarket Selection Committee that would provide recommendations to the City Council. With the help of a new $10,000 HUD grant, we hired a specialized market consultant and conducted an overall market study identifying three sites and five types of operators and then set out to determine the costs of operations for each site and for each type of operator. The market study also matched the identified sites with the types of potential operations and then determined how much each type of operator could expect to sell at each site given a certain size store. The type of potential operators included a conventional supermarket, a limited assortment supermarket, the relocation of the existing Onion River Co-op, and the relocation of the Onion River Co-op with either a conventional or limited assortment supermarket.

CEDO engaged a local design firm to develop several site scenarios for the redevelopment of the City-owned South Winooski Avenue lot. The site scenarios were crucial to determine costs for the redevelopment of the site as well as to determine how to best meet the needs of prospective supermarket operators.

We also conducted a citizen survey to determine citizen preferences regarding a supermarket for downtown. In addition, CEDO held public forums, contacted similar size cities to discuss experiences and opportunities, and contacted area brokers and other small supermarket operators to determine if there was interest. CEDO also sent out a notice to communities that receive HUD funding to request assistance in locating or developing a supermarket in our downtown.

This market study was supported with field research conducted by CEDO. We took a road trip seeking out grocery store models and prospecting for supermarket operators who might locate in Burlington. We wanted to develop a store that met all the food and sundry needs for the residents and that would meet all possible current and future trends based on our research and assessment of what was currently available in other communities. We visited supermarkets throughout Vermont and the Hanover Food Co-op in New Hampshire. Each store was videotaped, and the price of goods was tracked. We picked up store flyers to show the prices the stores were charging and documented every aspect of this process. Mindful of the controversy surrounding the project, we wanted to be thorough in our research. (In fact, it was later claimed that the price comparison surveys we developed weren't accurate. The claim that our survey was flawed was dropped when we explained that we collected fliers and took videos of the prices on items.)

On one of our stops on the road trip, we met with the manager for the Hanover Food Co-op. It had been in business since 1936 and consisted of two supermarkets and one corner store, with a range of conventional items along with more upscale items. It seemed to us that this store model would work for Burlington. We attributed the success of its operations to the following features: its made-to-order sandwich bar was top-notch; its pre-cut fruit was fresh, affordable and a good profit center; its prices were good, product selection was varied and its store layout was easy to navigate. It was also owned by community residents so there would be a good chance it would never move out of their neighborhood. We asked if the Hanover Food Co-op's management would advise the operator in Burlington so we could learn from their experience, and the Hanover Food Co-op later ended up providing consulting services to City Market which helped to ensure its success.

Based on the market study, citizen survey, supermarket site plan studies and field research, the City chose a City-owned site close to downtown that was a "brownfield" and over 100 years ago had been a ravine. This site included a municipal surface parking lot and adjacent parcel that was formerly a municipal police station that was deemed historic. The Police Department needed to move from the site as the building was too small for their needs and was falling apart. The site was also designated a "brownfield site" in need of remediation and ongoing monitoring. Originally, the site had been a ravine that was filled in a long time ago. Unfortunately, when we chose the site, the ground was caving in and was in need of shoring up. (The site was also rumored to hold a weapons cache from a long-ago war, but no evidence of such was uncovered in the development process. The lack of evidence was a great relief as it could have held up the entire project.)

Involvement of public stakeholders

While all this research was ongoing, the City also established a Supermarket Selection Committee with 15 members. Only one member was part of the City administration; the rest were citizens and city councilors. The citizen members represented the five neighborhood planning assemblies, City commissions, and interest groups.

The City issued a Request For Proposals (RFP) with the market study findings. Five groups submitted applications to meet the food needs of the downtown areas. We were very relieved we had supermarkets interested in the project because of the funds we had to return in 1989. The Committee reviewed the proposals, asked questions of the operators, held public hearings, and was charged with recommending two or three supermarket operators to the City Council and Mayor Peter Clavelle in accordance with the guiding principles we decided to adopt.

The list of guiding principles was created as a result of our research on the supermarket industry. We had met with the head of a regional supermarket chain and we implored them to carry more local goods. From these discussions we came to understand some of the industry barriers to providing local foods. For example, we learned that supermarkets charge rent or slotting fees to manufacturers or vendors to make room for a product on its store shelves.

During our research, we also discovered that the regional chains prefer to have vendors ship their goods directly to a central warehouse so they can consolidate the items to be shipped to their retail locations located in nearby states. Most of these warehouses are not located in Vermont. Both the slotting fees and the long distances to central warehouses were significant barriers for smaller vendors wanting to sell their goods through regional or statewide supermarkets.

This information was of great concern since our economic development strategy was based on local ownership and the development of sustainable agricultural products. We decided to develop a set of guiding principles that would assist us developing a supermarket project and came up with 26 different principles. Because the City owned land that could be used for a supermarket site, we had some leverage to ensure that at least some of these guiding principles could be included in any project that we decided to support.

The Committee held four meetings transcribed verbatim by a college intern. The City also hired the University of Vermont's Center for Rural Studies to conduct a study on consumer attitudes to a downtown supermarket. We ensured that the homeless population was surveyed by sending the surveyors to places where the homeless could be interviewed. The idea is that we wanted to build a store all residents wanted. The study showed that 83 percent of the participants preferred the store to be locally owned and operated. There was no point in building a supermarket that didn't meet local needs.

Meeting interim needs of the community

In the midst of all the research and development activities and before the new supermarket opened, the last downtown grocery store closed and CEDO was confronted with the need to ensure that Burlington residents, particularly the elderly, disabled, and low-income residents, had access to affordable food. To meet the needs of all residents, CEDO organized a committee that put in place the following action plan:

- Emergency funds were acquired and agreements were made for free taxi service and van transportation for elderly and disabled residents.
- Onion River Co-op offered home delivery services.
- Grocery store chains on the outskirts of town paid for expanded shuttle service to their stores.
- Thermal coolers and luggage racks were installed in the buses.
- 200 free bus passes were provided to those with special needs.
- CEDO published and distributed 5,000 copies four different times of a Resource Guide to food resources.
- CEDO developed a leaflet and distributed it to all the area residents, and posters were placed on the city buses and around town describing access options.
- CEDO staff encouraged a frozen food delivery service to expand with new trucks by marketing and promoting their service.
- Two Vermont Anti-Hunger Corps members were funded to work at the local food shelf due to the increase in households using this service.
- The City paid a nationally recognized grocery storey market research firm to evaluate all 18 corner stores in town. A meeting was conducted with the corner stores to hear from the consultants on how to improve the performance of their stores.
- Several corner stores added shelf space for groceries.
- A new bus shelter was installed by a traditional supermarket on the outskirts of town.

- The pharmacy that leased the site of the former downtown grocery store agreed to carry additional grocery items.
- Onion River Co-op and the local pharmacy that replaced the old grocery store agreed to host food drop-off boxes for a local food shelf.
- The City paid for the installation of refrigerators at the three "mini stores" opened in elderly housing facilities.

Selection of the supermarket operator

The Supermarket Selection Committee had five operator options to choose from those that applied. One was the local co-op, one was a small Vermont-based grocery store chain and one was a large supermarket chain that manages stores all over the Northeast and was owned by a firm from Europe. Two proposals, one from a discount supermarket chain and one from a small Vermont-based supermarket operator, were eliminated early on in the process because they did not express a strong commitment to work with the City in satisfying the goals and desires of the City in a timely and flexible manner. After contentious community debate, the Committee eventually chose the Onion River Co-op and a small Vermont supermarket chain as their two choices. The Co-op would develop a 32,000-square-foot supermarket and preserve a historic structure, and the Vermont-based operator would build an 18,000–25,000-square-foot supermarket. The choice was based on 39 criteria, including financial information, the proposed store format and other information helpful in evaluating the companies.

The Selection Committee strongly supported the concept of a public market (indoor farmers' market) within the store and thought it would complement either one of these supermarket operators. They declined to select the larger international supermarket chain as a potential operator based on the fact that the City met with their developer who insisted on wanting to build a bigger supermarket than we heard the neighbors wanted in their community. Additionally, the Supermarket Selection Committee believed that the development of a 50,000-square-foot supermarket at the downtown site would be unlikely and unachievable because it would have required a zoning change, required a large public subsidy for a parking garage, and several other factors that would have had big impact on the neighborhood.

There were also other obstacles to choosing the large supermarket operator. First, the neighbors that abutted the property threatened to go to court to block the project, which could have delayed the opening of the store by several years. Second, the large supermarket operator insisted that if it was chosen, and the store failed, the City should not be able to require that another supermarket be allowed to fulfill the lease. The City offered to operate the supermarket in case the large supermarket operator failed, but it refused, saying it wanted to control who would subsequently lease the space. This option, then, was not in line with the stated goal of ensuring that a supermarket be available now and in the future to serve our downtown. The choice became clear: to support the community-based, local food co-op and the small, Vermont-based supermarket chain.

Public controversy

The decision to support the Onion River Co-op proposal was fraught with controversy from political enemies, average citizens and Co-op members from the very beginning of the process. We were simply supporting the concept of moving a small, neighborhood-based co-op into a bigger, more modern home in the heart of downtown. The store would have 50 percent more space than the shuttered supermarket, sell popular traditional products as well as natural and organic foods and would not require membership to shop there. It didn't seem to matter what the facts were – a good portion of the public was opposed.

The controversy was intense and required our constant attention for over a year. There was nonstop news coverage and editorials about this project for three years. Almost all the coverage was unfavorable to what the City was proposing. In addition to the local press, the Public Broadcasting Service (PBS) ran a story that aired nationally that featured the controversy over this project.

The City conducted extensive public outreach to explain the decision to choose Onion River Co-op. A board member of the Co-op and former city councilor and I went to a senior housing complex with the PBS film crew. We held a public meeting with the residents who were incensed with the decision. The film was rolling and we were being cursed out by an elderly resident claiming City Market wouldn't sell meat, that they would be forced to eat certain types of foods and that the Co-op wouldn't meet their needs (luckily the expletives were deleted). We listened and explained that they could save money by buying in bulk because they could buy only what they needed, that they wouldn't need to buy and take home unnecessary packaging, that the Co-op would cater to their needs by carrying items they wanted to purchase, and that it would even deliver groceries directly to their door. Regardless of what we said, they didn't buy it and the controversy continued to simmer.

The large supermarket operator was incensed and hired more than 20 people to create petitions to be signed by Burlington residents calling for a special election. It wanted the City to bond for $800,000 to help pay for an underground parking structure. About 62 percent voted for this general obligation bond. But because the City Charter required 66.6 percent approval for a bond, the bond failed to pass.

Throughout the controversy, Mayor Peter Clavelle remained firm in his belief that local ownership was a key feature of sustainability. He said that we had been trying to promote local ownership and co-ops for years and this was our chance to do so. He stood firm behind the choice of the Selection Committee and advocated that the City Council approve the choice as well. During this period the Vermont-based chain decided it was no longer interested in operating a downtown supermarket in Burlington. Onion River Co-op was now left as the only preferred operator of the site.

Obtaining project financing and expertise

When Onion River Co-op (ORC) submitted their proposal to the City it included the Burlington City Land Trust (BCLT) as a partner. The idea was that ORC would provide half of the store with general merchandise and BCLT would develop an indoor farmers' market in the other half of the store (the public market.) At the time, ORC operated a small 3,500-square-foot storefront located in the Old North End neighborhood of Burlington. The ORC eventually decided to part company with BCLT. Based on the market research and our tour of supermarkets at the beginning of the process, we urged the Co-op to build a much bigger store, which it eventually agreed to do.

Once ORC became the preferred operator, it immediately conducted a market study that obtained almost identical results to the City's market study. The firm the City hired did a very thorough job. Actual sales were off from projections by only $60,000 at the end of the first year of the new City

Market Co-op's operations. One of the important lessons we learned was that independent market research is critical to help businesses build financial projections; it gives them confidence and helps get everyone comfortable with a project. City Market was able to exceed these expectations, which helped to quell the simmering opposition that was present even after it opened. According to the Food Marketing Institute's industry averages for comparable natural food stores, sales should average $493 per square foot, per year. The local co-op was achieving $1,366 per square foot annually before City Market even opened. City Market is now achieving over $2,000 per square foot in annual sales.

Due to Onion River Co-op–City Market's limited retained earnings, the City had to become involved with raising funds for Onion River Co-op–City Market to develop the project. As required by the City's selection process criteria, the Co-op had obtained and submitted a letter of intent with their proposal from a local bank to finance their project. When the bank was approached to finance the construction it declined. The City scrambled to underwrite a $2 million HUD Section108 construction loan which HUD finally approved that included 60 different loan-closing documents to be signed. To help with the project, the City hired an outside business consultant based in Brattleboro, Vermont to review the business plan and projections. We figured that he wouldn't be biased by all the news about this project, being located 115 miles south of Burlington. This was a concern at the time, as almost all the news about the project was sensationalized, which served to further polarize the community.

The success of this project depended upon many, many community partners. Just in terms of financing there were at least six entities involved, as follows:

1 *The City guaranteed the $2 million Section 108 HUD construction loan.*
2 *The Preservation Trust of Vermont raised $200,000 to help restore and repurpose part of the former police station that was deemed historic.*
3 *Vermont's senior Senator, Patrick J. Leahy, secured an HUD Economic Development Initiative earmark for $600,000 to pay for brownfield cleanup, the shoring-up of the ravine and preparing the site for construction.*
4 *A well-known Burlington resident, philanthropist and Vermont supermarket developer, donated equipment to the Co-op from one of his stores that was not being used.*
5 *The City provided $56,000 in grant funds from its Community Development Block Grant funds; and*
6 *City Market also raised $600,000 in equity from members and other city residents.*

The City also worked with City Market to find an architect who knew about supermarkets. It turned out this was not an easy thing to do. Most supermarkets have their own in-house architects and this is not a skill they let the general market provide. They control this flow of information. There is a science to building and operating supermarkets and this information is closely held. We identified lighting experts; same issue here too. We worked together to interview contractors. Onion River Co-op eventually selected a relatively new construction company. The construction company worked closely with the Co-op and its development consultant. The architect found a way to incorporate the old historic police station into the layout and design of the Co-op. The "newer" section of the former police department building was torn down and these materials were recycled. The new supermarket turned into an architectural student's dream – inside the building by the bulk food section you can see how a portion of the old historic building is incorporated into the new supermarket building. In the interior of the supermarket is the historic section of the old police station; it is dark, without natural light, and a perfect place to sell the bulk goods. You can communicate through the old windows between the back end of the produce and retail section.

The benefits of the supermarket project were laid out in a 2001 case statement by the Preservation Trust of Vermont and Onion River Co-op in order to raise money for its development. This is an excerpt from the case statement:

This project, which sees Burlington moving against the national trend toward massive corporate-owned supermarkets in suburban malls, will have many benefits for Vermont and the people of the Burlington area. The Onion River Co-op City Market will:

- Serve the residents of downtown and surrounding neighborhoods with a mix of conventional, natural, and cost-competitive food.
- Be an anchor food store for the entire downtown shopping area. This anchor, combined with several of the other large non-food anchors, will create a vibrant full service, one-stop shopping venue downtown that serves both local residents and visitors.
- Be locally owned and operated, which is important for community stability and sustainability in this age of impersonal national stores delivering local services and the resulting abandonment of many of those services when absentee ownership decides to pull out of a community.
- Strengthen the entire economic base of Burlington by attracting and retaining shoppers not only from Burlington but also from a significantly expanded market area. This will import dollars into the Burlington economy.
- Create a national model of how communities can work together to bring to fruition a project that will support locally owned business, combat sprawl, and significantly strengthen a threatened downtown core.
- Preserve and restore an historic building and put it back into productive use.
- Pay out wages of more than $2 million annually. The Co-op has committed to make every effort to hire residents of the neighborhood. Wages will average $8 to $10 per hour and at least 25 percent of the jobs will be full-time. Full-time employees will have a full benefit package.

Impact of planning commission requirements

The City's Planning Commission required that City Market build the structure fronted on the street instead of situated in the back of the parking lot like most supermarkets operate. They required that there be two floors on the section that fronted on the main street so it would blend in with the streetscape. Supermarkets are usually one big box, with one floor. So what's the big deal adding a second storey? It requires two means of egress, an elevator, extra space for hallways, the load that the ceilings take requires much stronger construction – read more money, much more money. So there is less space dedicated to retail on the ground floor and more "dead space."

This Planning Commission requirement of adding 3,000 square feet of second-storey space cost more money. To remedy the cost issues here, the City agreed to put in an additional $300,000 from the Capital Budget and the CEDO office agreed to move their Center for Community and Neighborhoods (CCAN) into the second-storey space for five years. The idea was that, at the end of 5 years, City Market would have the option to buy out the lease and CEDO would move into a new space. In 2007, CEDO moved into a new space and the Co-op took over the second floor as planned.

One of the important things we learned was the concept of " value engineering," i.e. how to cut back on expenses during construction, since the costs go up and the budget can't pay for everything needed. City Market could only borrow a certain amount and this was based on the capital they had, a ratio of debt to equity, and what the available projected cash flow would support. Also the HUD Section 108 program has collateral coverage requirements that limit the amount that can be borrowed. Financial and construction budgets kept being rewritten and items kept being taken out. The City was financing the project, held a lease on the land and building, and was responsible for making this project succeed. The pressure was constant and intense. We couldn't afford to move forward without eliminating budgeted items.

One high cost item we could eliminate was the freight elevator. The plan was to inventory goods on the second floor of the new market and the goods would be carried up and down the freight elevator. Without the elevator employees would have to hand-carry everything up a flight of stairs! The City Market staff successfully managed this issue in the short term by putting inventory on top of displays and employing the concept of "just in time" deliveries. Fortunately, over the years everything was put back in the budget as funding allowed.

Federal collateral requirements

One big pressure was the collateral requirement for HUD's Section 108 program. When a deal is underwritten and collateral is required everything including the kitchen sink is required for collateral. CEDO had to pledge our future annual Federal Community Development Block Grant (CDBG) allocations for five years as additional collateral. If the project failed and the building collateral wasn't sufficient, we would have to use future CDBG grants that funded all of our office operations, future neighborhood projects funds, and future capacity grants for community organizations. We were betting the farm on the project.

Opening and managing the new City Market

Once financing was secured and construction began, the Co-op management was then faced with the daunting task of opening a new store that embraced the 26 identified principles. Management had to start new departments including a meat and fish department, deli, limited service restaurant, prepared foods, and flowers. They had to hire and train a whole new staff and capitalize and create systems they didn't have. They had to compete with supermarket chains that centralize to spread the cost among many retail outlets, all while trying to pay the staff a livable wage. It had to sell at least 1,000 Vermont products, necessitating more staff to receive goods and handle the vendors. They were also required to sell a series of basic items at fair prices which met the needs of low-income, elderly, and disabled residents.

City Market operated at a loss for three years and struggled to be a store with competitive prices that catered to everyone living in the neighborhood. This was not an easy task and something City Market struggled with. After City Market had opened and been operating for a few years the City graded the Co-op a "C minus" at meeting the needs of low- and moderate-income residents – a low grade for our most important priority.

Fortunately, City Market took this feedback to heart and was determined to improve. It decided to sell to people that were on food stamps and other government supports at a 10 percent discount. This program would be supported by the store's higher-income customers. About a year of lobbying, with the help of Vermont's congressional delegation, was spent trying to secure the U.S. Department of Agriculture's (USDA) permission to do so. But the USDA said it would be unfair to all the customers to do this, so it wasn't allowed. The Department said it would be unfair to charge low-income customers a 10 percent reduction in their food purchases.

As an alternative, City Market decided to offer the 10 percent discount anyway to combat childhood hunger. Customers had the option to join at no cost (a hardship waiver) but were encouraged to join the Co-op to be able to vote and accrue towards a Patronage Refund. The number of low-income customers continued to grow over time. The program was made permanent, and now 1,110 of the Onion River Co-op's 8,439 members are low-income. The criteria are: to be receipt of Social Security disability benefits, Women, Infants, and Children (WIC) or Supplemental Nutrition Assistance Program (SNAP) benefits. USDA later held a national conference and invited City Market's general manager to speak. He relayed this saga and next thing you know City Market was the cover story on USDA's trade journal.

The 26 guiding principles

As stated previously, we developed a set of 26 guiding principles that any supermarket operator would be required to agree to if it was located on City-owned land. These principles included a requirement to meet the needs of low-income, disabled and elderly residents; specified operating hours; a guarantee that the building would always be a supermarket; a commitment to hire from local neighborhoods; a commitment to pay livable wages, and to sell a minimum of 20,000 household items with 1,000 Vermont-made or locally-sourced products.

Two of these principles are highlighted below to demonstrate how they helped to shape the development and implementation of the project.

Principle 1: requirement to carry Vermont-made products

One of these principles required that the supermarket operator carry at least 1,000 Vermont-made products. We wanted to promote locally produced food as specified in the City's 1994 strategic economic agricultural policy, and we knew producers needed a market. We also knew from working with the supermarkets in our community that there were high barriers to entry for small businesses, such as the previously mentioned slotting fees. When we brought this issue to one regional supermarket store owner he said this requirement was impossible to meet. Ten years later in 2010, this same supermarket chain hung a banner over its front door announcing that it sold 1,000 Vermont products. (Note that the City Market–Onion River Co-op began carrying 1,700 Vermont-made products five years after they opened!)

Principle 2: green materials and construction practices

Another important principle was the use of green materials and construction practices, incorporating features such as the use of natural lighting. At this time, the U.S. Green Building Council's Leadership in Energy and Environmental Design (LEED) sustainability standards were being developed and we were looking into best practices for new construction. LEED is a complex rating system for the design, construction, and operation of high-performance green buildings.

We followed the LEED standards when we designed and built the supermarket, but didn't apply to be rated by the system. We felt a couple of the criteria were not consistent with the needs of a downtown supermarket. First, they required more parking spots than were then practical for the size of our site, and second, they required two electric car charging stations. Based on the use of other electric charging stations in the city, we felt two were unnecessary and would be difficult to site. We also tried to be innovative in other ways, which led to complications. For instance we wanted to have skylights for natural lighting. We found a chain of supermarkets that had some skylights and a study confirming that the chain with the skylights sold more food per square foot.

City Market installed 16 skylights, but areas of the produce section heated up and lights that were supposed to dim with natural light streaming from the skylights didn't work properly. Eventually the building systems were debugged and the building now functions as it was designed. Because of the importance of keeping environmental standards high, we went to great lengths to inform the public about the green building standards that were being implemented. The store also received an award for the green nature of the project from the State of Vermont. Over the years as capital budgets allowed, more green building improvements have continually been made to the building.

Building a green supply chain

By Clem Nilan, former general manager, City Market

There is a certain infrastructure necessary to make a local sales program happen. Supermarkets shy away from this commitment because of the higher labor costs associated with the endeavor. Instead of simply ordering from a warehouse, each outlet needs to develop their own sourcing network. This requires enhanced buying activities and a much more complex receiving procedure, as well as scenarios for dealing with supply challenges.

The benefits of selling local foods to the community are very powerful. Rather than money leaving the community to buy goods, money recirculates in the community. This so-called multiplier effect for local agriculture (2.5) is one of the highest of any sector of the economy according to by Chuck Ross, Vermont's secretary of agriculture. The multiplier effect is the idea that money spent locally to buy Vermont products creates additional income for a Vermont farmer or producer, who spends part of this money boosting the income of yet another person, and so on. It is a powerful economic driver of local economic growth.

A significant challenge facing us was in tracking local sales, many of these products do not have a UPC code. Yet our store is driven by data. There is an absolute need to quantify the amount of local food we sell. We had to develop some in-house POS procedures to enable this to happen.

On the positive side we are told repeatedly that our customers come to us because they know that they can find local food in our co-op; selling local foods is a strong sales driver. Nationally, co-ops differentiate themselves in the marketplace through their commitment to sell local food. Selling local food builds strong community connections. Our members and customers can often find the same food in the Co-op that they can at the farmers' market. They know that we are supporting our neighbors. They realize that the food isn't traveling from thousands of miles away. Some cite the cost of local food as a deterrent. Here at City Market there is virtually no pushback on the price of local food.

To summarize, selling local food is the major point of differentiation between us and the competition in the marketplace.

A model for other communities

In 2007 the City hosted a New England conference of the National Community Development Association. We helped organize the conference for our peers who run community development departments in the bigger cities of the Northeast like Boston, Hartford, Providence, and Portland. We were suggesting topics for the conference and offered the idea of the "Food Sector as an Economic Development Strategy," with a case study on the development of City Market. The conference organizing committee didn't like the idea because it didn't seem relevant to the audience's work. We stuck to our guns and organized a workshop on this effort.

It turned out that, just before the conference, the same large supermarket operator that had responded to the City of Burlington's RFP, closed a supermarket in Connecticut and two inner-city supermarkets in Rhode Island. East Providence's community development director, who had helped organize the conference, went from asking: "Why do you want to host this workshop?" to, "Our mayor wants me to figure out what to do because our supermarket is pulling out." It turned out that the company that had proposed building a supermarket in Burlington had developed several supermarkets in the Northeast during the time we were developing our supermarket, and had later decided to close three of them.

In 2006 a surprise party was held for Mayor Peter Clavelle. He was retiring from the mayorship after 15 years (the longest-serving mayor in Burlington's history), and the party was held in the same room for the public meeting where the PBS crew had filmed the city councilor and myself years before at the Senior Residence Building. There were survey results from residents posted on the bulletin board that asked what residents' favorite activities were. The number one activity was the free monthly lunches that City Market brought over to the building. We finally felt vindicated.

Politics with that, anyone?

Things got a little crazy. The choice between a conventional grocery store and a member-owned co-operative became emotionally and politically charged. The local media ran a series of articles as public hearings were held, and the issue became rather contentious. Meg Klepeck, City Market's local foods coordinator, provides an overview of what happened:

> It was portrayed as a bit of a class war in the Public Broadcasting Service (PBS) documentary *People Like US: Social Class in America*. A petition called for a citywide referendum providing the City with the ability to offer a large supermarket operator an $800,000 tax subsidy. It was defeated and City Council selected the Onion River Co-op proposal. The City added a lease addendum to address resident concerns that arose during this process.

Get more info about the documentary at: www.pbs.org/peoplelikeus/film/index.html.

Behind the scenes: getting to a yes vote on the downtown co-op

By George Thabault, former assistant to Mayor Bernie Sanders and Mayor Peter Clavelle

The City administration risked a lot of political and financial capital when it decided to stand behind the concept. We immediately faced strong opposition to the idea from all sides including Co-op members. Some opponents viewed the multinational chain as a much better alternative and pressed for a citywide referendum on the matter. The company itself boasted that it had "more experience, greater financial resources," and that it would "generate more tax revenue for the city." A special election on the referendum fell short of a required two-thirds majority for passage – but not by much. The close vote put more pressure on the City's negotiations with the Onion River Co-op, and a complex development agreement headed for the City Council for approval. Amidst the constant political battles and back-and-forth claims and counterclaims in the press and among citizens, the Council's approval was not a sure thing.

Sometimes direct lobbying is not the wisest course of action. Personal pressure tends to freeze people into firm positions early in the process. It can be more helpful to create a positive climate for elected officials considering a controversial vote. If you can create an air of reasonableness about an impending "tough vote," you have a better chance at winning the day than if you harangue individuals with declarations about how "right" your view is.

We fostered that positive climate with a solid, down-to-earth letter to the City Council in support of the co-op project. The letter was signed by people many in town would consider "reasonable" and fair-minded. We also sought the support of a small grocery store owner. Lou Merola had been a co-owner, along with several of his seven brothers and four sisters, of "Merola's Store," a small but successful family enterprise that had occupied several locations in the city. With over 40 years at the counter, he probably knew more voters personally than all the Council members combined. Lou was semiretired and a little skeptical of the co-op project. Still, he lived only a couple of blocks from the supermarket that had left town and understood that a lot of the profit margin would leave town if a large chain succeeded in landing the project. He decided to risk his reputation as a savvy business grocer and sign the letter.

The Council approved the economic development agreement with the Co-op, even as the local newspaper railed and predicted fiasco at every opportunity. After the Council meeting, one of the opposing councilors walked by me, leaned in and whispered, with a trace of admiration, I think, "How in the hell did you get Lou Merola to sign that letter?" I knew then that we had succeeded in creating an open space, a little room for a "soft landing" as the Council made the tough decision to select the Co-op.

Looking back

The controversy surrounding the project was the most challenging aspect of it. In the process we learned that you can't take yourself too seriously and you can't be concerned about people that talk to you about "the project of the day." Local politics is a local sport and it is part of the job. It brings out the worst in people and it brings out the best in the projects. After working under a microscope for over 28 years we learned that it is best to give as much information about a project to the public as possible so that they can make an informed decision – knowing that there is wisdom in the crowd even in the midst of possible opposition to sound leadership. That doesn't mean you don't stand your ground – it just means that you also have to listen and take other voices into account.

Success through commitment and hard work

The development of City Market began with a commitment from the leadership of the City of Burlington. It was a commitment to serve all the citizens of Burlington by retaining a downtown supermarket with principles that coincided with a long-term sustainable economic development plan. The goal of securing access to local foods for all was achieved through strong partnerships, a lot of time, investment, perseverance, and hard work for the greater good.

The City Market–Onion River Co-op was chosen by the Selection Committee and the Burlington City Council to develop the market because of its experience, expertise, and commitment to the people of Burlington. They worked hard over the years and have exceeded all expectations. The benefits accrued are high, and City Market is a valued asset for the community. City Market is now the highest-grossing single-store grocery co-op in the United States with annual sales of $34,000,000 per year. Total patronage dividends have stayed within our community, and from fiscal year 2008 to 2012 total almost $2,000,000, with an average annual patronage dividend check of $80. City Market has paid $235,000 annually to the City of Burlington for local taxes and rent for the land. They have offered 149 classes with 1,547 people attending annually.

City Market is powered by the sun

Solar project stats

Grocery stores use large amounts of electricity, so the return on investment, in dollar terms, isn't high. Using 136 rooftop panels from groSolar of White River Junction provides 31.28 kw; this solar-generated electricity produces the equivalent of enough electricity to power over six Burlington homes. The solar project cost $187,912 with some of the cost picked up by a $53,900 grant from The Vermont Clean Energy Development Fund of the American Recovery and Reinvestment Act. The Co-op also received a $40,204 federal tax credit (30 percent of the Co-op's investment in the project.) This has brought out-of-pocket costs for the project to $93,808 and a return on investment of over 13 percent, and a payback of the costs in less than five years.

Solar projects for the community

In addition to the new solar panels, City Market and groSolar have partnered to offer a free "Solar Made Simple" seminar each month, to help educate local residents and business owners about the new economics of solar power.

(Source: www.citymarket.co-op/SolarPower)

Resources and ideas for making it happen in your community:

Unemployment data for the U.S.: www.bls.gov/news.release/metro.t01.htm

See the booklet, "How Burlington Became an Award Winning City": www.cedoburlington.org/business/CEDO percent2027 percent20Years.pdf

A comprehensive guide for employees and employees: www.cedoburlington.org/business/chittenden_county_resource_guide/cc_resource_guide.htm

Guide – "Doing Business in Burlington": www.cedoburlington.org/business/doing_business_in_burlington/TheGuide.pdf

The 2010 Legacy Action Project Report Card: http://burlingtonlegacyproject.org/files/2010/03/reportcardprintable.pdf

Promote lingering around: Put a gum ball machine, free coffee and hot chocolate in your waiting room.

Get to know your elected officials: Meet with city, county, and state officials to share what you're doing, thinking about doing, or to develop joint proposals to move ideas into action.

Get to know appointed officials: Watch city or county commission meetings in areas you might be interested in like planning and zoning, public works, utilities commission, city council, transportation authorities, etc.

Management by wandering around: Go for walks and stop in to talk to business owners. Get to know local business people and their needs and concerns. Solve a major problem for each business, if possible.

Caring for your community: Take regular walks through the city and observe the small details. Let the appropriate people know if you see or hear about something that is not functioning properly, including potholes in the streets, sidewalks, curbs, green belts, street lights, cross walks, curb cuts, handicap ramps, directional signs, fencing, facades, truck loading zones, parking lots, on-street parking, common areas, and trash removal. All of these items need to be functioning well if you want your city or neighborhood to be a place where people want to raise their family. Don't expect that it is other people's responsibility and that they will take care of it. It is very satisfying as you see these issues get resolved over time. Look, listen, observe and then work to make things right.

How to change road construction nightmares into traffic that flows better than ever: If you are facing a road construction season that will cause regular traffic jams, establish a staffed regional committee to manage the traffic flow and set up weekly meetings. Include members of the police department, general contractor, public works department, government transportation planner, member of the business association, economic development staff, town managers' offices affected by the traffic. Develop a marketing and communications plan that includes regular and timely announcements sent to the local media for drive-time announcements. Erect street signs with a clever name to call, like 658-GOGO, and develop and distribute brochures with info on the construction plans. Develop strategies to inform the public using the internet. Make sure that someone from the committee talks to the flaggers at least weekly in the beginning to find out what and how they are doing. Provide a feedback loop to the contractors to make sure they are aware of problems that can and will exist. Ensure that the contractor or his/her representative speaks with every business that will be directly affected well in advance to give them warning, and to find out if they have special plans that might be impacted by construction or vice versa.

Develop a commercial space database: Create lists of developable land and vacant commercial and industrial spaces. These resources draw people to your office so you can find out how else you can help them.

Life long learning: Learn how to develop cash flow projections and read and dissect financial statements. Take a course on construction management and facilitative leadership.

Appendix A

City of Burlington: Honors, Accolades and Articles

June 1988	Tied for first place as "Most Livable City" by U.S. Conference of Mayors for populations under 100,000.
June 1991	Voted "Best in the Northeast" by *Inc.* magazine in its "Best Cities for a Growing Business."
June 1993	Featured in the book *50 Fabulous Places to Raise Your Family* by Lee and Saralee Rosenberg.
June 1993	Rated best place in the nation for raising children by the group *Zero Population Growth*, Washington, D.C.
November 1993	Listed in a survey entitled "Top 10 Best Cities for Running a Home-Based Business" in *Home Office Computing* magazine.
June 1995	Ranked second by *Zero Population Growth* of the ten best places in the nation to raise children.
June 1995	Topped list of seven "Dream Towns" by *Outside* magazine.
September 1995	One of five cities listed in the *New York Times Sunday Magazine* in "The Rise of the College City: The Best New Place to Live."
September 1995	Featured in the book *A Good Place to Live* by Terry Pindell.
February 1996	Cited as one of the seven best retirement areas by *New Choices, Living Even Better After 50* magazine.

June 1996	Ranked as the seventh-hippest arts town in *100 Best Small Art Towns in America* by John Villani (1996, John Muir Publications).
September 1996	One of "The 100 Best Small Art Towns In America" (#7) by John Villani (*The 100 Best Small Art Towns in America*).
April 1997	Ranked sixth best "family-friendly" place in the nation by *Reader's Digest*.
April 1997	Burlington's Church Street Marketplace is one of five national winners of the Great American Main Street Award from the National Trust for Historic Preservation.
April 1997	Named one of "10 Great Places to Raise a Family" by *Parenting* magazine.
April 1997	One of the "25 Most Livable Cities in America" (with populations under 100,000) by U. S. Conference of Mayors.
May 1997	Ranked #4 of "America's 10 Most Enlightened Towns" by *Utne Reader* magazine.
May 1997	"Burlington: Northern Light" (*Nation* magazine).
May 1997	Named one of four outstanding getaway locations in Northeast by *USA Weekend*.
November 1997	Ranked #3 of "Best Cities in U.S. for Women" by *Ladies Home Journal*.
November 1997	Named "A Latte Town" by the *Weekly Standard* ("The Rise of the Latte Town").
May 1998	Cited "One of 15 Best Walking Cities in America" by *Walking* magazine.
October 1998	Ranked #4 "Best Beauty Spots" by *Ladies Home Journal*.

August 1998	*Bus Tours Magazine*: "Burlington, Vermont."
January 1999	Burlington named #5 "Boomtown: 75 Top Cities to Start a Business" (behind Seattle, Austin, Las Vegas, and Denver) by *Point of View* magazine.
May 1999	USAirways *Attache*: "Navigating Through Burlington."
June 1999	Ranked #1 for "Families That Love Outdoor Sports" by *Outdoor Explorer* magazine, premier issue.
October 1999	One of ten "College Towns Worth A Visit" – *Princeton Review*, "The Best Colleges."
November 1999	Ranked #1 of "Top Ten Cities To Have It All" by *Arts & Entertainment TV*. Runners-up: Chapel Hill, NC and Austin, TX.
March 2000	One of America's "Ten Fittest Cities for Women" – *Health Magazine*.
May 2000	One of "50 Best Places to Live" – *Maturity Magazine*.
Summer 2000	*Rails to Trails*: "Vermont's Holy Grail."
November 2000	Burlington named #7 "Healthiest Place for Women to Live" by *Self* magazine.
Aug 2001	Ranked #1 "Kid-Friendly" smaller city (100,000–2 million metro) for quality of life by *Zero Population Growth*.
May 2002	"Hip, Hippie, Hippest": *Washington Post*, 5/26/02.
September 2002	"36 Hours – Burlington, VT," *New York Times*, 9/20/02.
October 2002	Burlington named #2 "Happiest," and #4 "Healthiest" city for women by *Self* magazine.

April 2003	Burlington named "One of the 50 Best Places to Live" and "Editor's Pick in the Northeast" by *Men's Journal* magazine.
April 2003	Burlington named "One of America's Dozen Distinctive Destinations" by the National Trust for Historic Preservation.
May 2003	Featured in "Secrets of Successful Cities" in the *Hartford Courant*.
May 2003	Featured in *Men's Journal*, "The 50 Best Places to Live."
June 2003	Burlington named "One of the Five Best Places to Live and Ride" by *Bike* magazine in its June 2003 issue.
March 2004	Burlington named #7 small city in the country for doing business by *Inc.* magazine.
April 2004	Burlington is one of five U.S. cities to receive *Delicious Living* magazine's "Impressive City Award" for its "exceptional efforts toward sustainable living."
April 2005	Ranked as the "third-funkiest city in the world" by British Airways' magazine *Highlife*.
June 2005	Burlington named #12 among the top 25 small cities in *American Style* magazine's "Top Arts Destinations" for 2005.
Nov 2005	Burlington included in *50 Fabulous Gay-Friendly Places to Live*, by Gregory A. Kompes, published by Career Press.
April 2006	Featured in the "50 Best Places to Live" list of *Men's Journal* as one of five "Best of the Best" cities with "the perfect combination of adventure, attractiveness, and affordability."

March 2007	"Burlington, VT is nation's most eco-friendly city" – Newsday.com by John Curran, 3/7/07.
January 2008	"Vermont Weekends, Burlington, VT," by Molly Walsh, *Ski* magazine.
October 2008	"Small City, Big Charms," by Daniel Machalab, *Wall Street Journal*, 10/18/08.
October 2008	Church Street Marketplace named one of top ten "Great Public Spaces in America" by the American Planning Association.
November 2008	"CDC: Burlington, VT is Healthiest City," by Mike Stobbe, Associated Press, 11/18/2008.
June 2009	"Hill Section One of Best Neighborhoods 2009" – *This Old House*.
March 2010	"One of Best Cities for New Jobs this Spring" – *Forbes. Com*
April 2010	"First Wave City" – *Carbon War Room*.
May 2010	"Prettiest Town in America" – *Forbes.Com*
May 2010	"Tree City USA" – Arbor Day Foundation.
May 2010	Featured in "Top 100 Places to Live in America" by *RelocateAmerica.com*
June 2010	"Top 10 City for the Next Decade" – *Kiplinger's Personal Finance* magazine.
June 2010	"#1 Bass Fishing Capital" – *Outdoor Life*.
Summer 2010	Featured in "Top 25 Cities for Art (Small city category)" by *American Style* magazine.
September 2010	"Planning Innovation to Institutionalize Sustainability Award" – *ICLEI*.

October 2010	"'Innovative' Sustainability Plan Selected as Noteworthy Program and Role Model" – Harvard Kennedy School of Government, Ash Center.
December 2010	Award of Excellence for Sustainable Community Development from the National League of Cities.
Summer 2011	Featured in "Top 25 Cities for Art (Small city category)" by *American Style* magazine
Summer 2012	Burlington ranked "#1 Place for Guys" by *Men's Health* magazine.
December 2012	Vermont named "Healthiest State in the US" by the United Health Foundation.

Appendix B

Thirty Year Timeline and Milestones for the City of Burlington's Community and Economic Development Office (CEDO)

1983

- CEDO founded.
- Applied for Urban Development Action Grants (UDAG):
  - The Maltex Building: $675,000 helped restore the vacant 40,000 sq. ft. building on Pine Street which was completed in 1985 and later incubated Dealer.com and Lake Champlain Chocolates.
  - The Wells-Richardson Building: $300,000 helped renovate the 18,300 sq. ft. building on College Street which was completed in 1984.
  - The Holloway Block Project: $215,000 in grants helped revamp this block near Battery and Main Streets which contained some of the oldest buildings in Burlington, creating 6,000 sq. ft. of office space.
  - Worked with FM Burlington, the City's largest property tax payer, on $25 million plans to construct a major department store, expand Burlington Square Mall and the Radisson Hotel (now Hilton), and provide parking with a $4 million UDAG to be completed in 1985.

1984

- E-Z Access Program (grants to businesses for wheelchair modifications) launched.
- Burlington Revolving Loan Program for business begins awarding loans through CEDO.
- Supported Alden Waterfront Project for mixed-use development of public open space, boathouse, bike path, museum, children's center, visitor information, and housing. Also enabled citizen participation and tracked public opinion.

- Created Old North End Redevelopment strategy offering over 80 small business and housing improvement loans and grants and supported a new Community Police Program.
- Published *Jobs & People: A Strategic Analysis of the Greater Burlington Economy*, a 25-year blueprint for CEDO's economic development programs.
- CEDO funding launched the non-profit Burlington Youth Employment Program (BYEP) providing training for disadvantaged youth.

1985

- The Vermont Energy Investment Corporation, established to help businesses, landlords and tenants meet energy-efficiency goals, was originally formed with CEDO staff and funding assistance from BHA and VHFA. VEIC currently employs 275 people.
- South End Arts + Business Association formed with financial and technical assistance from CEDO.
- Received a $1.6 million Urban Development Action Grant distributed as loans to the Maltex Building on Pine Street for incubator space as well as the downtown Park Plaza and Wells-Richardson Building projects.
- City of Burlington establishes Tax Increment Financing District to spur development and help pay for future public improvements on the waterfront.

1986

- Step Up for Women (now called Vermont Works for Women), created and funded by CEDO, is a program that trains women for non-traditional careers that pay a livable wage.
- City mandated that 10 percent of the construction jobs for publicly funded construction projects over $50,000 be held by women.
- Completed the $25 million FM Burlington project downtown.
- Burlington Local Ownership Project, created out of the recommendations of the original *Jobs & People* report, gave preference to locally and employee-owned businesses and pursued

job creation strategies for women, minorities, youth, and people from low-income backgrounds.

- Gardener's Supply locates in Burlington and begins a community-wide effort to clean up and revitalize the Intervale with technical assistance from CEDO. Gardener's Supply is now employee-owned and currently employs over 250 people.
- CEDO organized the "Downtown Summit" to enhance the viability of the central business district by promoting the Marketplace through capital improvements, fairer fees and more parking availability.

CEDO awards

- "Certificate of National Merit" awarded to CEDO for the Village at Northshore and Howe Meadow – by the U.S. Department of Housing and Urban Development for its model efforts as a public/private partnership.

1987

- CEDO and the Chamber of Commerce launched the Burlington Vermont Convention Bureau with a $25,000 city appropriation.
- CEDO received $1.6 million Urban Development Action Grant (UDAG) to assist in the financing of a downtown supermarket.
- CEDO conceived of and was a major sponsor, along with the Small Business Development Center and the State of Vermont, of the Vermont Innovation Summit, helping local companies obtain Small Business Innovation Research Grants.
- Developed Vermont Products Innovation Network with UVM.
- Transportation and Parking Council, staffed by CEDO, created to develop recommendations to improve parking supply and management.
- The long vacant Strong Lot on Main Street was developed by Ray Pecor with CEDO support, into Courthouse Plaza, a six-storey office building with three decks of parking.
- CEDO-drafted state legislation provided legal certainty to worker co-ops and financial incentives for incubators.

CEDO awards

- National Award for Outstanding Leadership in Citizen Volunteerism, U.S. Conference of Mayors.
- National Award for Outstanding Leadership in Development for Burlington's local ownership development efforts – U.S. Conference of Mayors.

1988

- CEDO managed development of the Community Boathouse and 7.5-mile bike path along Burlington's waterfront.
- Women's Small Business Program established in collaboration with Trinity College and supported annually with CDBG funds.
- Received $2 million Urban Development Action Grant (UDAG) to develop the Corporate Plaza Project, developing an 80,000 sq. ft. office building, creating a 325-space parking garage, 4 units of housing and retaining Key Bank downtown.
- CEDO developed Minority Business Assistance Program using a $75,000 U.S. Small Business Administration Grant.
- Assisted in redevelopment of the 105,900 sq. ft. incubator space at the former Lane Press headquarters, now called the Kilburn and Gates Building, located on Pine Street.
- Created Chittenden County Roundtable, allowing for discussion between economic development managers, town managers, and planning and zoning directors.
- Greg Noonan's 1986 book, *Brewing Lager Beer*, helped break open the microbrewing industry in the U.S. In 1988, Noonan and his partner, Steve Polawacyk, opened Vermont's first brew pub called Vermont Pub & Brewery in Burlington with CEDO support.
- CEDO sponsored conference on Export Marketing leads six local businesses to begin exporting to Canada.

CEDO awards

- Citation award for the Burlington Urban Design Study – *Progressive Architecture* Magazine.

1989

- *Jobs & People II* commissioned by CEDO and presented to the Governor's Commission on the Economic Future of Vermont.
- Burlington Community Banking Council created and staffed by CEDO.
- CEDO commissioned a Credit Needs Assessment of the community to determine the extent of unmet credit needs and evaluate banks' performance with regard to criteria established by the Community Reinvestment Act. Assessment was completed by UVM professor and city councilor Jane Knodell.
- Worked with UVM's Church Street Center in developing dozens of low cost courses for the business community.
- CEDO, the City Attorney, neighboring and statewide organizations helped convince state regulators of the potential adverse social, environmental, and economic impacts of the proposed Pyramid Mall in Williston. After a 17-year effort, Pyramid Mall agreed not to build a department store greater than 50,000 sq. ft. and to build a series of buildings around a town square instead of a typical mall.

CEDO awards

- Excellence on the Waterfront Honor Award – The Waterfront Center.
- National League of Cities and Towns, Innovations Award for Job Training and Development, Women's Economic Opportunity Program.

1990

- Waterfront Revitalization Plan drafted and approved by voters.
- $11.2 million Energy Conservation Bond issued.
- Worked with Merrill Lynch to relocate within downtown Burlington and with Onion River Co-op to expand to a larger facility in the Old North End.
- CEDO conceptualized a project to renovate 294 North Winooski Ave. (former Fassetts Bakery Building) creating 25,000 sq. ft. of incubator space for small businesses, and helped obtain VEDA (Vermont Economic Development Authority) financing. This project housed

Microstrain, a business that originated through research at UVM and now employs 50 people.

- CEDO helped organize four annual Alternative Career Forums with area colleges and Vermont Student Assistance Center. Over 600 people attended annually with a goal of encouraging people to choose careers in the non-profit and socially-responsible business fields.
- CEDO report on the local impacts of military cutbacks leads to the creation of a Manufacturing Task Force with the Chamber to attract new manufacturing jobs.

CEDO awards

- Special Appreciation Award – Downtown Burlington Development Association.

1991

- Waterfront Urban Reserve purchased and Waterfront Park created.
- Relocation of Naval Reserve began with commitment of $2.5 million of federal funding.
- Linked Deposit Program established, awarding city accounts to local banks on behalf of their community reinvestment activities.
- Recycle North founded and supported annually with CDBG funding.
- CEDO-drafted mandatory recycling ordinance passed by City Council resulting in over $400,000 in state grants for capital equipment.
- Vermont Businesses for Social Responsibility established with CEDO staff assistance and funding.
- In cooperation with the Downtown Burlington Development Association and the Church Street Marketplace, CEDO organized and staffed the Downtown Partnership which developed comprehensive recommendations to improve downtown.

CEDO awards

- Finalist – Innovations in State and Local Government: Housing Programs – Ford Foundation/JFK School of Government.

1992

- CEDO's VISTA Program initiated, bringing 20–40 staff annually to community organizations.
- CEDO's Microenterprise Program launched, providing information, referral, technical business assistance, loan program services to 300 very small businesses annually.
- 69,000 sq. ft. Burton Snowboards world headquarters and 27,000 sq. ft. Rhino Foods operation moved to Burlington Industrial Park (in former buildings of General Electric – a defense contractor) with financial and technical assistance from CEDO and Greater Burlington Industrial Corporation.
- Worked with the State of Vermont to locate the Department of Health Headquarters in a new 110,000 sq. ft. building in downtown, creating 200 construction jobs and maintaining 200 Health Department workers.
- Convinced State to locate a new 80,000 sq. ft. Chittenden County Courthouse adjacent to present courthouse downtown.
- Provided Frog Hollow financial and technical assistance in locating new store on Church Street, offering 15 new Burlington craftspeople a retail outlet.
- Co-sponsored conference with Downtown Business Development Association on Americans with Disabilities Act.
- Worked with Parks Department to secure a grant from the Small Business Administration to plant new trees in downtown.

1993

- Supported the new $2 million Coast Guard Station and installed 30 boat slips.
- Developed and distributed a New Business Location package including a new logo for economic development.
- CEDO funding helped restore a King Street neighborhood grocery and provided a training grant to Recycle North to provide employment skills to homeless people.
- Residential Improvement Program completed ten grants in the Old North End. The Facade Improvement Program also provided funding to help restore commercial properties on lower North Street.

- Secured long-term lease for and renovations to City Hall Park to accommodate and expanded Farmers' Market. This is currently Vermont's largest farmers' market.

CEDO awards

- Citation Award – Northern New England Tradeswomen.

1994

- Former J. C. Penney's building rehabilitated into a combination 63,000 sq. ft. downtown bookstore and office building with CEDO's financial and technical assistance.
- Secured funding to close top block of Church Street to automobile traffic and to renovate and expand the Church Street Marketplace.
- Howard Bank moved 60 employees back to downtown from Williston.
- State Courthouse opened on Cherry Street with CEDO support.
- Vermont Expos baseball team brought to Burlington through CEDO, downtown business, and community efforts.
- Peer lending established, loaning $160,000 to 68 businesses.
- *Jobs & People III: Towards a Sustainable Future* published.
- Vermont Downtown Program conceived by and created with CEDO support, providing grants, loans, and tax credits to eligible downtown projects statewide.
- CEDO provided grant to Old North End Credit Union (now called Opportunities Credit Union) which enabled them to move to a larger facility.
- CEDO helped develop the Climate Action Plan for the City.

1995

- Burlington Neighborhood Project created, providing grants for neighborhood-initiated community projects.
- Enterprise Community Designation established; $3 million to the Old North End with over 70 projects funded.
- Citywide AmeriCorps*VISTA Program created.

- Magic Hat Brewery opened in South End with financial and technical support from CEDO.
- Vermont Sustainable Jobs Fund, conceived by and developed with CEDO and Vermont Businesses for Social Responsibility support, receives a $250,000 annual appropriation from the Vermont State Legislature.
- Worked with Key Bank and UVM Business School to establish business community program.
- CEDO funded the pre-development of a new building in the South End on Briggs Street which is now the home of Petra Cliffs Climbing Center.

1996

- Brownfields Program launched by CEDO with a $200,000 EPA grant.
- Pratt & Whitney 40,000 sq. ft. building and 15,400 sq. ft. Aviatron building allowed companies to locate in Airport Industrial Park with financial assistance from BCDC (Burlington Community Development Corporation) and technical support through CEDO and the Airport.
- Surveyed 500 local businesses, with Saint Michael's College, to better understand how economic conditions and public policies impact them.
- Began collaborative technical assistance for the Micro-enterprise program to target specific neighborhoods and market sectors.
- Worked with the Lake Champlain Chamber of Commerce to establish the School to Work Initiative, securing $87,000 in funding to begin and operate the program. The program is now known as navigate and operates throughout Vermont.
- Collaborated with Burlington Electric, UVM, the Intervale Center, and Gardener's Supply to develop a plan for an Eco-Industrial Park near the McNeil Station, a 50-megawatt wood chip-fired power plant.
- Good News Garage, a program providing affordable vehicles to individuals living in poverty who need reliable transportation, began with CEDO assistance.

1997

- Time of Sale Energy Ordinance passed, requiring new owners of multi-unit residential properties to bring their buildings up to energy-efficiency standards within 12 months.
- Rose Street Artist Co-op (12 units), a former bakery, developed by the Burlington Community Land Trust in the Old North End with financial and technical support from CEDO including funding from the Enterprise Community program and general city funds.
- $5.4 million Section 108 loan guarantee funds secured from HUD for economic development, affordable housing, and public infrastructure.
- Assisted Vermont Transit Lines in their relocation to a vacant building on Pine Street.
- Assisted in expansion of Gregory's Supply and All Season Kitchens.
- Vermont Probation and Parole offices were relocated to Burlington.
- Using HUD Section 108 funds, CEDO provided $1 million to reconstruct Lake Street, upgrade shower facilities at the Boathouse, and stabilize the vacant Moran Plant located on the Waterfront.

1998

- Filene's (now Macys) parking garage project secured $1 million EDA grant funds as well as $2.5 million in State Downtown funds.
- Secured approval for a $1.02 million grant from the Economic Development Administration for the Riverside Eco-Industrial Park.
- Collaborated with local employers, Lake Champlain Workforce Investment Board, state departments, and Recycle North to design and submit a comprehensive welfare-to-work service sector training proposal.
- Supported redevelopment of former Hood Building into mixed-use facility on South Winooski Avenue.

CEDO awards

- Best Practice Award: CDBG Advisory Board – HUD.

1999

- Vacant Buildings Ordinance passed, providing incentives to develop vacant buildings. Since this ordinance, the number of vacant buildings has decreased from 39 to 30.
- Park Place Co-op built for $5.2 million, creating 34 units and 17,000 sq. ft. of commercial space in downtown with CEDO's financial and technical support.
- Community Outreach Partnership Project (with UVM, CEDO, and other organizations) created for three years to bolster local economy with $1.4 million grant secured from HUD.
- Lake Champlain Chocolates expands into 25,000 sq. ft. building at 750 Pine Street with technical and financial support from CEDO.
- Business Refugee Resource Guide published by CEDO.
- Champlain Valley Network created, a website providing information about entry-level jobs, opportunities for advancement, and other resources collaborating with Vermont Department of Employment and Training, Lake Champlain Region Workforce Investment Board, and Cyberskills Vermont.
- Associates in Rural Development, the largest international consulting firm based in Vermont, expands onto Church Street with CEDO assistance. The staff has grown from 65 to over 120.
- CEDO loaned $400,000 in a CDBG Interim Float Loan to Burlington Community Land Trust for the redevelopment of the former Vermont Transit Bus Barns into a mixed-use redevelopment at the gateway to the Old North End.

CEDO awards

- Best Practice Award McClure Multi-Generational Center – HUD.

2000

- CEDO completed new Consolidated Economic Plan.
- Filene's 150,000 sq. ft. Department Store opened.
- Specialty Filaments retained on Pine Street (200 manufacturing jobs).
- Fresh Connections (home meal replacement manufacturer) stabilized and expanded with CEDO financial and technical assistance.

- Select Design (marketing services company) relocates in South End in 60,000 sq. ft. building with technical support from CEDO.
- CEDO provided financial and technical support for the transition to and creation of 32,000 sq. ft. City Market downtown grocery store for $3.9 million.
- Enabled expansion and retention of architectural salvage warehouse on Main Street.

CEDO awards

- Semi-finalist, Innovation in State and Local Government: Community Justice Center – JFK School of Government.

2001

- $30 million Burlington Town Center Mall renovation, attracting or expanding 23 retail tenants and creating 300 downtown jobs with the support of CEDO and other City departments.
- Ethan Allen Shopping Center completely redeveloped, with CEDO support, for $6 million, adding a new 50,000 sq. ft. Hannaford supermarket, post office, and a new commercial building in the New North End.
- City received Vermont Downtown Historic District designation and used it to obtain $1 million in Historic Tax Credits to renovate 20,000 sq. ft. Hall Block Commercial Building.
- CEDO published *Guide to Doing Business in Burlington*, which won an international award.
- General Dynamics planned to redevelop their facility into the Innovation Center of Vermont, retaining 500 jobs in 160,000 sq. ft. of space. The building was renovated over a five-year period.
- CEDO targeted vacant upper story commercial properties in downtown for restoration and secured $75,000 in state funds to develop six units of housing in two buildings.

2002

- Renewal Community designation received, providing federal tax incentives for business development in the Old North End and downtown.
- City Market opened downtown, creating 107 new jobs, providing the City of Burlington $167,717 in taxes and fees in 2007 and $225,526 in cash donation to the Food Shelf since 2005.
- Blodgett Ovens was retained in town, preserving 200 livable-wage jobs.
- Church Street Marketplace renovations begun with CEDO-assisted $1.8 million federal grant.

CEDO awards

- Community Impact Award: Burlington Truancy Task Force – United Way of Chittenden County.
- Best Practice Award: AmeriCorps*VISTA Program – Corporation for National Service.
- Burlington Neighborhood Project – National Community Policing Partnership Award – Met Life.

2003

- General Dynamics was retained; 160,000 sq. ft. business converted to design and testing center. Began five-year, $50 million restoration of building located in the South End with GBIC and CEDO support.
- Burlington Telecom's Fiber Optic Network completed for all 41 city buildings and sites, with CEDO assistance.
- The CEDO-staffed BCDC (Burlington Community Development Corporation) purchased land for $1.8 million with HUD Section 108 Program in the Urban Renewal area for development of hotel and parking garage.
- CEDO formed Community Coalition with the IRS and community partners to improve access to free tax assistance and credit counseling.
- Formed Microbusiness Alliance to foster collaboration among community business assistance providers and worked with

Microbusiness Development Program at the Champlain Valley Office of Economic Opportunity (CVOEO).

- Rewrote Consolidated Plan for Housing and Community Development.
- Community Health Center Dental clinic in Old North End opened with a $300,000 CEDO float loan, leveraging a $700,000 federal grant in meeting a critical need for affordable dental care.

CEDO awards

- *The Guide to Doing Business*, Best Special Purpose Publication – The International Economic Development Council.
- Best Practice Award: A*VISTA – National League of Cities and Towns.

2004

- Vermont Software Developers Alliance created with CEDO financial and technical support.
- Burlington hosted 500 attendees of the Sustainable Communities International Conference, originally conceived by Mayor Peter Clavelle, with financial and technical support from CEDO.
- Lake Street Extension rebuilt with $495,000 in HUD Section 108 loan underwritten by CEDO.
- Marsh Management retained downtown (Vermont's largest captive insurance company with 80 employees) with CEDO technical support.
- Burlington's "Creative Economy" was recognized by an article in the *Harvard Business Review* as the fourth best in the nation.
- CEDO began smart growth initiative offering detailed information with redevelopment opportunities in the South End of Burlington.
- Began construction on North Street revitalization project.
- Opened Center for Community and Neighborhoods (CCAN) which offered funding, support for Neighborhood Planning Assemblies (NPAs), and Neighborhood Development Grants. The Community Justice Center offers Restorative Justice panels and Graffiti First Response Teams.

CEDO awards

- Community Impact Award: Study Circles on Racism – United Way of Chittenden County.

2005

- Secured $6 million in new federal funding for infrastructure renovations on the Church Street Marketplace.
- CEDO worked with Koffee Kup Bakery to allow them to expand at their current location and remain in Burlington.
- CEDO helped Charlebois Truck Parts obtain $1 million Commercial Revitalization Deduction to support 15,000 sq. ft. addition.
- CEDO established BE3 (Burlington Energy, Environment, and Economy) project to help restaurants and convenience stores become more effective in using resources.
- Airport received state and city permits to expand the Industrial Park. CEDO secured $4 million, the largest VEDA (Vermont Economic Development Authority) industrial loan in its history to construct a new, 40,000 sq. ft. facility for Heritage Flight.
- $13.5 million, 113,000 sq. ft. Lake and College Project opens on Waterfront with CEDO support. Seventh Generation, a company which makes environmentally friendly cleaning products, moved its world headquarters to the building in 2006 with location help from CEDO and is now LEED (Leadership in Energy and Environmental Design Green Building Rating System) certified.
- Burlington Telecom business service begins, providing affordable internet, telephone, and television. By 2008, 140 businesses are using this service.

CEDO awards

- The *Guide to Doing Business*, Best and Companion Disc, Superior Award – Northeast Economic Developers Association.
- Best Practice Award: Increasing Access to Affordable Housing – HUD.
- Award: First Response Team – Keep America Beautiful.

2006

- Filled next to last remaining vacant parcel in the City's Urban Renewal area with a 116-room hotel, creating 41 full-time jobs, expanding the Lakeview parking garage with 450 new spaces, and creating 31 units of housing. CEDO provided financial and technical support for this project.
- CEDO worked with Specialty Filaments to sell their 105,000 SF property to Lake Champlain Chocolates and Dealer.com.
- Lake Champlain Chocolates renovated their portion of the building, which is now the first LEED-registered warehouse and distribution project in Vermont with CEDO support.
- Dealer.com established three workforce training programs with CEDO support from Vermont HITEC (Vermont Healthcare and Information Technology Education Center), training 29 new employees. The company, currently employing over 200 people, completed a $5 million renovation and received LEED certification with CEDO financial and technical support.
- Completed a 66,000 SF, $6.6 million North Street Revitalization Project in the heart of the Old North End.

CEDO awards

- Best Practice Award: Cities United for Science Progress, Lead Safe ... for Kids' Sake.

2007

- CEDO hosted and helped organize a Federal Reserve Bank of Boston public meeting on strategies for local food systems, the creative economy, and affordable housing in Burlington.
- Vermont Agency of Human Services located downtown.
- Awarded a Preserve America grant from the National Park Service to develop a web-based guide to Burlington's Cultural and Historic Resources promoting heritage tourism and an American Battlefield Protection Program grant to build a War of 1812 memorial in Battery Park.

- CEDO supported formation of the Old North End Arts and Business Network.
- Worked with the new managing partner of the Burlington Town Center to increase visibility through a $2 million "facelift" on the Church Street entrance and helped facilitate the public planning process on improvements to Church Street side streets.

CEDO awards

- Welcome to the Neighborhood Award – Vermont Housing and Conservation Board.

2008

- Burton bought an 84,000 sq. ft. building on Industrial Parkway with $1.4 million in state incentives and technical support from CEDO and GBIC. They plan to expand their world headquarters in Burlington.
- Four-story, 15,000 sq. ft. vacant Hinds Lofts restored downtown, with CEDO support, for $2.5 million, adding 15 residential units.
- Dealer.com adds 70 new employees; company now employs 205 workers.
- Community-based research group, with CEDO and BED support, began examining the use of waste heat from Burlington Electric Department's McNeil Plant to heat and cool homes and businesses.
- Hyundai and Subaru car dealerships renovated and rebuilt new locations at gateway to Burlington with CEDO technical support.
- CEDO helped conceive of and organize two Vermont 3.0 Creative Tech Career Jams at Lake and College on the Waterfront and at Champlain College. Over 2,500 people attended the free events with 50 exhibitors and 12 lectures.
- CEDO, working with UVM, helped organize alliance of businesses in the BioScience industry.
- Rewrote Consolidated Plan for Housing and Community Development.
- CarShare Vermont established with CEDO support.

- CEDO convened meeting with 14 business leaders and seven business technical assistance providers to determine status of the local economy.

2009

- Heritage Aviation deconstructed then completely renovated and expanded an 80,000 sq. ft. former Army National Guard Hanger into a LEED-certified, state-of-the-art General Aviation Facility. The company serves both the commercial and general aviation communities with fueling and deicing services, general Fixed Based Operator (FBO) as well as aircraft maintenance and avionics services.
- Terry Precision Bicycles chose to locate in Burlington and rented two locations in the South End totaling 10,000 square feet. they employ 17 people, including 13 in Burlington.
- Construction is underway on Phase 1 of College Street Waterfront Access Project. Improvements include new decking at the Boathouse, upgrading of the path to the Boathouse, sheet pile retaining walls and new sidewalk on Lake Street between Main and College Streets.
- Moan Plant redevelopment project, on the waterfront, CEDO applied for and received federal funding totaling over $3 million dollars.
- Transformed the Gosse Court Armory building into a community-wide asset and demonstrated Burlington's commitment to sustainable practices through adaptive reuse of a community and recreation center. CEDO leveraged $63,000 from EPA and $20,000 for clean-up activities from the Vermont Army National Guard and over $1.2 million for redevelopment through fundraising, private donors, and public dollars.

CEDO awards

- Excellence in Architecture Design Awards for Moran Center on the Waterfront – Vermont Chapter of the American Institute of Architects.
- People's Choice Award for Moran Center on the Waterfront in Burlington – Vermont Chapter of the American Institute of Architects.

- Brownfields Success Story for Gosse Court Armory – US Environmental Protection Agency.
- Cities Untied for Science Progress Award for Burlington Lead Program – The United States Conference of Mayors.

2010

- Dealer.com decided to expand in Burlington renting additional office space in three locations in the South End and hiring 100 new employees with a total of 400 employees.
- The Courtyard Burlington Harbor has expanded, adding 16,000 square feet of building space, which equates to an additional 34 guest rooms. This brings their downtown hotel capacity to 161 rooms. The hotel hosts approximately 55,000 vistors to Burlington annually.
- Situated on Battery Street, a 45,000 sq. ft., "A" grade, multi-tenant office building with a two-level parking structure has completed the permit approval process. Construction and occupancy is estimated to be in the spring of 2012.
- CEDO worked with the Vermont BioScience Alliance, Vermont Legislature, the Vermont Center for Emerging Technology and GBIC to secure $4 million in state funding for a new Seed Capital Fund for growth companies.

CEDO awards

- Stakeholder Partnerships, Education and Communication Award for Volunteer Income Tax Advice and Tax Counseling for the Elderly – from the Internal Revenue Service (IRS).
- Brownfield Success Story for 13 properties owned by Champlain Housing Trust (CHT) in the Old North End – U.S. Environmental Protection Agency (EPA).
- Engaged Community Partner Award for meaningful and committed engagement with your higher education partner for mutually beneficial results – The Vermont Campus Compact in 2010.

2011

- Fleet Fuel Reduction workshop organized with University of Vermont Transportation Research Center. Twenty-five people representing fleets based in Vermont attended . The City agreed to open its fast-fill natural gas station to cab companies.
- Hotel Vermont began construction. Last property in the Urban Renewal area to be developed closing the 40-year effort to redevelop this site. The hotel was supported with a Section 108, $1.8 million loan and Department of Public Works providing the funds for a parking garage.
- Cornell Trading Company opened a retail store on Battery Street adjacent to their world headquarters.
- Series of meetings held to scope out future activity to support the creative economy.
- College Street Waterfront Access project completed. $2.2 million in improvements undertaken including reconstruction of the Pease parking lot to accommodate tour buses and school bus drop off and pick up, reconstruction of the circle at the bottom of College Street to accommodate public transit, the reconstruction of College Street from Battery Street to the Boathouse to improve pedestrian and vehicle flow, new public restroom, and information area.

2012

- Thayer School project in the New North End began with CEDO financial and technical support. Champlain Housing Trust, Cathedral Square and a private developer are undertaking a mixed-use and mixed-income development on the former DMV/Thayer School site. Project includes mixed-income apartments for 69 senior households, apartments for 33 low- and moderate-income families and another 85 market rate units.
- ICV built with CEDO technical support Waterfront Plaza which is situated within walking distance to Lake Champlain. This 45,000 square-foot, "A" grade, multi-tenant office building with a two-level parking structure located in downtown Burlington has been completed and is 90 percent occupied.

- Lake Champlain Chocolates purchases building on Pine Street and is planning an expansion.
- Burton Snowboards continues to upgrade their building and grounds on Industrial Parkway.
- Switchback Beer starts bottling operation to coincide with their tenth anniversary. Demand outstrips supply.
- Vermont Technology Alliance moves headquarters to Burlington.
- Dealer.com finishes upgrading the second half of their building and their employment grows to 675 employees. Renovations to the new space, included construction of a mezzanine and expansion of office space. Vermont Economic Development Authority approved a $6.4 million tax-exempt Recovery Zone Facility Bond, purchased by People's United Bank, and a $1.3 million Direct Loan to help with the $13 million expansion project.

Notes

1 The quest for a durable local economy

1 *http://moneywatch.bnet.com/economic-news/blog/daily-money/the-10-happiest-and-saddest-cities-in-the-us/2308/* or see the Gallup US Well-Being index.

2 www.resilientus.org/library/CARRI_Definitions_Dec_2009_1262802355.pdf

3 Stockholm Resilience Center, www.stockholmresilience.org/research/whatisresilience.4.aeea46911a3127427980004249.html

4 www.billmckibben.com/deep-economy.html

5 www.aae.wisc.edu/pubs/cenews/docs/ce269.txt

6 *The Fourth Sector Network (*2009). *The Emerging Fourth Sector*, Washington, D.C.: The Aspen Institute. www.fourthsector.net

7 John Abrams (2008). *Companies We Keep*, White River Junction, VT: Chelsea Green Publishing Company.

8 Ibid.

9 See the E. F. Schumacher Society at www.smallisbeautiful.org/about.html, and its new organization, http://neweconomicsinstitute.org/

10 E. F. Schumacher (1973). *Small Is Beautiful: Economics as if People Mattered*, London: Harper & Row, p. 191.

11 See the New Economics Institute, http://neweconomicsinstitute.org/

12 See any of McKibben's work, including the 1987 classic, *The End of Nature*, or more recent work such as *Deep Economy*: McKibben, B. (2006). *The End of Nature*, New York: Random House. McKibben, B. (2007). *Deep Economy: Economics as if the World Mattered*, Oxford: Oneworld Publications.

13 Frances Moore Lappe and Anna Lappe (2003). *Hope's Edge: The Next Diet for a Small Planet*, New York: Penguin.

14 See for example, *The Small-Mart Revolution*, and BALLE, the organization he helped co-found, www.livingeconomies.org/

15 www.neweconomics.org/content/history-nef

16 See *Jobs & People* reports on the CEDO website, www.cedoburlington.org
17 William Murtagh uses this expression in relation to community historic preservation efforts. See book, *Keeping Time*, New York: John Wiley & Sons, 1997.
18 www.cedo.ci.burlington.vt.us/

2 Localize and socialize

1 Carr, J. H. and Servon, L. (2009). "Vernacular Culture and Urban Economic Development," *Journal of the American Planning Association*, 75(1): 29.
2 Ibid., 30.
3 Shuman, Michael (2007). *The Small-Mart Revolution: How Local Businesses Are Beating the Global Competition*. San Francisco: Berrett-Koehler Publishers, Inc., pp. 46–50.
4 Phillips, Rhonda (2002). *Concept Marketing for Communities*, Westport, CT: Praeger Press.
5 *Bloomberg Business Week*, "The New Entrepreneur. Maryland Passes 'Benefit Corp. Law for Social Entrepreneurs,'" Posted by: John Tozzi on April 13, 2010. www.businessweek.com/smallbiz/running_small_business/archives/2010/04/benefit_corp_bi.html
6 See information from www.livingeconomies.org/
7 John Abrams (2008). *Companies We Keep: Employee Ownership and the Business of Community and Place*, White River Junction, VT: Chelsea Green.
8 Daniel Barlow, "Sanders: Worker-owned businesses can save our economy," *Rutland Herald*, August 27, 2010.
9 Alison Teague, "Cooperating in commerce," commonnews.org/site/site02/story.php, October 24, 2010.
10 Abrams, *Companies We Keep*.
11 Internal City of Burlington memo, CEDO, "Burlington Local Ownership Development Project," 1984.

3 Welcome one, welcome all: economic inclusiveness

1 Vaclav Havel, quoted in John Abrams (2008). *Companies We Keep*, White River Junction, VT: Chelsea Green Publishing Company, p. xii.

2 Abrams, *Companies We Keep*, p. xii.

3 See *Jobs & People 2010* at www.cedoburlington.org

4 See the article, "Examining the connection between women, employment, and incarceration in Vermont," by Tiffany Bluemle, in *Women, Girls and Criminal Justice*, June/July 2008, 9(4): 49–64.

5 See www.mercyconnections.org

6 See the guide at www.cedoburlington.org/business/refugee_resource_guide/busref_resguide_toc.htm

7 This case is excerpted from a report by Jim White and Anne Peter, *The Vermont Refugee Microenterprise Program*, City of Burlington, 2006.

4 Crunch and funk: cultural vibrancy

1 Jackson, as cited in J. H. Carr and L. Servon (2009). "Vernacular culture and urban economic development," *Journal of the American Planning Association*, 75(1): 28–40, at 29.

2 Cannavo, P. F. (2007). *The Working Landscape: Founding, Preservation, and the Politics of Place*, Cambridge, MA: MIT Press.

3 Loner, Michael C. (2004). "Working to develop sustainability within Burlington's creative community". June 8, 2004. White Paper, available at www.snellingcenter.org. www.snellingcenter.org/filemanager/download/3304/%20%20

4 Vermont Council on Rural Development, "Advancing Vermont's creative economy," Montpelier, VT, 2004.

5 "Burlington's South End: The SoHo of Vermont," *Vermont Business Magazine*, March 2003.

6 Richard Florida, www.creativeeconomy.org/

7 Stuart Rosenfeld, "Just clusters: Economic development strategies that reach more people and places," September, 2002, Regional Technology Strategies, Inc. http://rtsinc.org/publications/index.html#2002

8 Rosenfeld, "Just clusters."

9 See the City of Burlington's Municipal Development Plan at www.ci.burlington.vt.us
10 This case was developed by Alison Flint, CEDO research intern, summer 2009.
11 Carris, D. (1998). "Pine Street map and history," unpublished paper, SEABA archives, Burlington, VT.
12 Chittenden Historical Society in 1991: Blow, D. (1991). *Historic Guide To Burlington Neighborhoods*, Burlington, VT: Chittendon Historical Society, pp. 83–91.
13 DVD, Sanders/Lafayette Debate W/Pine St. Bus. Assoc. February 20, 1987, CCTV.
14 Karen Unsworth, daughter of Ray Unsworth and current Mapleworks and Howard Space owner, Correspondence, 2009.
15 Mark Waskow, SEABA board president, Correspondence, 2009.
16 *Vermont Life*, Spring 1982. www.vermontlifecatalog.com/
17 *Jobs & People*, 1984, p. v.
18 SEABA minutes: May 1, 1986.
19 *Jobs & People*, 1984.
20 New York: www.artsandbusiness-ny.org/; Chicago: www.artsbiz-chicago.org/; Philadelphia: www.artsandbusinessphila.org/
21 SEABA board meeting minutes, May 1, 1986.
22 "Free retail sales seminar for merchants and employees," Press Release, Bruce Seifer, CEDO and "The Greater Pine Street Business Association sponsoring their first seminar," Press Release, Tom O'Brien, Lincoln Works.
23 Correspondence between Gail Hanson at National Life and GPSBA Insurance Subcommittee.
24 SEABA board meeting minutes, August 1988.
25 *Pine Street Newsletter*, Autumn 1989.
26 *New York Times* article, December 27, 1987, "Portland concerns try day care partnership."
27 *Champlain Business Journal*, "South End Arts and Business Association is working vehicle for pride in community," September 2001, p. 23.
28 "Draft Supplemental Environmental Impact Statement – Southern Connector/Champlain Parkway." VTrans, October 2006. www.aot.

state.vt.us/SouthernConnectorSEIS/SouthernConnectorDraftSEIS TableOfContents.htm

29 DVD, Sanders/Lafayette Debate, W/Pine St. Bus. Assoc. February 20, 1987, CCTV.

30 SEABA special meeting minutes (re: Contract 6 of Champlain Parkway), July 15, 1993.

31 January, 1992, Correspondence, Letter to appropriate party from David Griffin, SEABA president.

32 *SEABA Newsletter*, "Pine Street News," Spring, 1992.

33 "EPA proposes cleanup plan for the Pine Street Canal Superfund site," November, 1992.

34 SEABA board meeting minutes, April 1, 1992

35 SEABA board meeting minutes, September 2, 1992.

36 Martin Feldman, quoted in "Superfund: unearthing a politically desirable solution," by Martin Feldman and Cara Robechek, unpublished paper, December 12, 2006.

37 Ibid.

38 *Evaluation report on the Pine Street Barge Canal Coordinating Council, Burlington, VT: Lessons learned from this Region 1 Community Advisory Group*, internal evaluation report, EPA Region 1, July 2000.

39 "Blanchard Beach fact sheet," City of Burlington, June 27, 2007. www.ci.burlington.vt.us/mayor/press_release/docs/Press%20 Releases/20070627Blanchard%20Beach%20Fact%20Sheet.pdf

40 New England Regional Water Program, "Business Friends of Engelsby Brook: Engaging the business community in NPS pollution prevention." www.usawaterquality.org/

41 New England Regional Water Program, "Business Friends of Engelsby Brook: Engaging the business community in NPS pollution prevention" www.usawaterquality.org/NewEngland/Topics/hot/ Engelsby_Brook.htm

42 Ibid.

43 Ibid.

44 Tess Taylor, Correspondence, 2009.

45 *SEABA Newsletter*, Spring 1992.

46 *SEABA Newsletter*, Fall/Winter 1992/1993.

47 Mark Waskow, Correspondence, 2009.

48 Mission Statement:
The South End Arts and Business Association enhances the economic vitality and eclectic mix of Burlington's arts and business community in the area south of Main Street by: *Promoting* our unique blend of art, commerce, industry and entrepreneurial spirit; *Providing* an influential voice to promote and ensure our member's common interests; *Informing* our members of issues that affect our community and the actions we take on its behalf.

49 Resolution 5.0. May 10, 2010, Councilors Shannon, Paul, and Adrian.

50 Acquino, J., Phillips, R., and Sung, H. (2013). "Tourism, culture and the creative industries: reviving distressed neighborhoods with arts-based community tourism," *Tourism, Culture and Communications*, 12(1): 5–18.

5 I'm OK, You're OK: social well-being

1 Rhonda Phillips and Robert Pittman (2009). *Introduction to Community Development*, London: Routledge, p. 6.

2 Ibid., p. 7.

3 Heerad Sabeti (2009). *The Emerging Fourth Sector*, Washington, D.C.: The Aspen Institute, p. 2.

4 Source: Richard Cohen and Tim Hansen-Turton, "The birth of the fourth sector," in *Philadelphia Social Innovations Journal*, October 2009, www.philasocialinnovations.org/site/index.php

5 *Jobs & People IV*, Update October 8, 2010. This report was made possible through the collaborative efforts of many people. Bruce Seifer coordinated the efforts to create this update. Nancy Brooks at Cornell University is responsible for the majority of the updated data in this report. Alison Flint, former CEDO intern, contributed analysis, organization, and new material.

6 Source: *Jobs & People IV*.

7 Speech by Peter Clavelle, Mayor, City of Burlington, Vermont. Maine Neighborhoods Conference, Portland, December 14, 2005, "Creating and sustaining livable neighborhoods."

8 This section is excerpted from the Legacy Plan, see http://burlingtonlegacyproject.org/projects/neighborhoods/ for more details as well as the plan document.

9 Because Burlington's unemployment rate is now among the lowest in the U.S., it no longer meets criteria to receive AmeriCorp volunteers.
10 See the Legacy Project and plan at www.cedo.ci.burlington.vt.us/legacy/
11 Burlington Legacy Project, www.cedo.ci.burlington.vt.us/legacy/
12 See more details and information at www.goodnewsgarage.org
13 www.goodnewsgarage.org/About-Us/Impact.aspx

6 Glowing and growing: energy and environment

1 From: https://www.burlingtonelectric.com/page.php?pid=141&name=Burlington%20POWER%20Program%20
2 From United States Environmental Protection Agency Greenhouse Gas Equivalencies Calculator, www.epa.gov/cleanenergy/energy-resources/calculator.html#results
3 See the following for a study of the waterfront redevelopment of a brownfields site: http://epa.gov/brownfields/sustain_plts/factsheets/moran.pdf
4 Source of this summary: www.epa.gov/region1/brownfields/success/burlington_vt.html
5 Source: www.epa.gov/ne/brownfields/success/10/Burlington_VT_Old_North_End.pdf
6 This article first appeared in the Federal Reserve Bank of Boston's winter 2009 issue of *Communities and Banking* magazine. The views expressed do not necessarily reflect those of the Federal Reserve Bank of Boston or the Federal Reserve System. www.bos.frb.org/commdev/c&b/2009/winter/Seifer_Vermont_Sustainable_Jobs.pdf
7 Written by Bruce Seifer, who has been instrumental in crafting the original concept for both the Vermont Businesses for Social Responsibility and the Vermont Sustainable Jobs Fund.
8 The author floated the original idea and included it in a Burlington, Vermont, economic development policy paper. Next it was presented to VBSR and discussed at their board, policy committee, and visioning meeting. The result was the Sustainable Jobs Coalition. Collaboration with a variety of organizations led to the passage of state legislation.
9 See www.vsjf.org/peer_collaborative/purpose.shtml

7 Tasting as good as it looks: local food system sustainability

1. Ben Hewitt (2009). *The Town that Food Saved: How One Community Found Vitality in Local Food*, Emmaus, PA: Rodale Press, p. 4; Barbara Kingsolver (2007). *Animal, Vegetable, Miracle: A Year of Food Life*, New York: Harper, p. 5.
2. "America's healthiest and unhealthiest states: States in New England perform best, while the South still falls behind." www.forbes.com/2010/12/06/healthiest-unhealthiest-states-lifestyle-health-uhc.html?partner=email
3. See the executive summary at: www.vsjf.org/news/37/farm-to-plate-executive-summary-released
4. National Gardening Association, http://assoc.garden.org/
5. National Gardening Association.
6. The information for this section is from the Legacy Project materials, for more information see http://burlingtonlegacyproject.org/
7. www.foodsecurity.org/FPC/
8. See the evaluation report for the original three-year funding cycle at http://crs.uvm.edu/evaluation/bsfp_execsumm06.pdf
9. From Burlington Food Council at http://burlingtonfoodcouncil.org/
10. Organic Trade Association's 2010 Organic Industry Survey, www.ota.com/organic/mt/business.html
11. www.uvm.edu/foodsystems/
12. Excerpted and reprinted with permission by the Wallace Center at Winrock International, www.winrock.org , "Community food enterprises, local success in a global marketplace," 2009. Thanks to Travis Marcotte, the Intervale director, for updates and additions.

8 Summing up: lessons learned and other insights

1. Vermont HITEC, www.VTHITEC.org

About the Authors

Rhonda Phillips: Community well-being is the focus of Rhonda's research and outreach activities. Author/editor of 15 books on community development and related topics, she offers both practice and academic perspectives on the ever changing topic of community revitalization as a professor at Arizona State University's School of Community Resources and Development.

Bruce Seifer is a consultant with deep experience in economic development. He led the City of Burlington, Vermont's economic development efforts for three decades, providing technical assistance to 4,000 businesses and numerous nonprofits. Bruce frequently speaks at national forums on policy and strategy, city revitalization, and program design and evaluation.

Ed Antczak: After 20 years in business, Ed joined CEDO's Economic Development Division in 2003, focusing on assisting businesses at all stages of growth, managing a revolving loan fund, and as a member of various development project teams. He currently serves on the steering committees of several national sustainable economic development organizations.

Index